Modern Cosmology

Modern Cosmology

Alicia Knight

MURPHY & MOORE
www.murphy-moorepublishing.com

Modern Cosmology
Alicia Knight
ISBN: 978-1-63987-370-8 (Hardback)

MURPHY & MOORE

Published by Murphy & Moore Publishing,
1 Rockefeller Plaza,
New York City, NY 10020, USA

Cataloging-in-Publication Data

Modern cosmology / Alicia Knight.
p. cm.
Includes bibliographical references and index.
ISBN 978-1-63987-370-8
1. Cosmology. 2. Earth (Planet). 3. Planets. 4. Astronomy. I. Knight, Alicia.
QB981 .M63 2022
523.1--dc23

For more information regarding Murphy & Moore Publishing and its products, please visit the publisher's website www.murphy-moorepublishing.com

Table of Contents

Preface

The branch of astronomy which deals with the origin of the universe is called cosmology. It focuses on the study of the constant evolution and eventual fate of the universe. The subject of cosmology is divided into physical cosmology, religious cosmology and philosophical cosmology. Physical cosmology makes use of mathematical principles to study the physical origin and evolution of the universe. Religious cosmology provides an explanation of the origin and evolution of the universe from religious perspectives. It may involve the description of a creator or a pantheon of deities. Philosophical cosmology deals with topics beyond scientific explanation. It is a multidisciplinary field of philosophy which draws on epistemology, metaphysics, philosophies of science and fundamental physics. This book provides comprehensive insights into the field of cosmology. It provides significant information of this discipline to help develop a good understanding of cosmology and related fields. This book is an essential guide for both academicians and those who wish to pursue this discipline further.

A detailed account of the significant topics covered in this book is provided below:

Chapter 1- All the existing matter, space and time collectively make the universe. Some of the concepts studied in relation to the universe are cosmic backgrounds, Hubble's law, Friedman equations, etc. This is an introductory chapter which will introduce briefly all the significant aspects of the universe including chronology, age and shape of the universe.

Chapter 2- Cosmology is a branch of astronomy concerned with the studies of the physical universe including origin and evolution of the universe. The chapter elaborates big bang theory, cosmological spacetime, structure formation, cosmological components and cosmological constant. This chapter has been carefully written to provide an easy understanding of the varied facets of cosmology.

Chapter 3- The theory of special relativity determines the relationship of space and time for objects moving in a straight line at a constant speed. The topics elaborated in this chapter are Lorentz transformations, tests of special relativity, general relativity, Einstein equation, etc. It discusses in detail the theories and methodologies related to special and general relativity.

Chapter 4- Physical cosmology is a branch of cosmology concerned with dynamics of the universe as well as the origin, structure, evolution and ultimate fate of the universe. This chapter discusses physical cosmology, Friedmann-Lemaitre-Robertson-Walker Metric, big-bang nucleosynthesis and absolute time and space. All the diverse principles of physical cosmology have been carefully analyzed in this chapter.

Chapter 5- Dark matter is composed of particles that are not detectable by observing electromagnetic radiations because they do not reflect, absorb or emit light. This chapter discusses in detail cold dark matter, hot dark matter, mixed dark matter and dark matter halo. These topics are crucial for a complete understanding of the subject.

I would like to make a special mention of my publisher who considered me worthy of this opportunity and also supported me throughout the process. I would also like to thank the editing team at the back-end who extended their help whenever required.

Alicia Knight

1

Universe: Past, Present and Future

All the existing matter, space and time collectively make the universe. Some of the concepts studied in relation to the universe are cosmic backgrounds, Hubble's law, Friedman equations, etc. This is an introductory chapter which will introduce briefly all the significant aspects of the universe including chronology, age and shape of the universe.

The universe is everything. It includes all of space, and all the matter and energy that space contains. It even includes time itself and, of course, it includes you. Earth and the Moon are part of the universe, as are the other planets and their many dozens of moons. Along with asteroids and comets, the planets orbit the Sun. The Sun is one among hundreds of billions of stars in the Milky Way galaxy, and most of those stars have their own planets, known as exoplanets.

The Milky Way is but one of billions of galaxies in the observable universe — all of them, including our own, are thought to have supermassive black holes at their centres. All the stars in all the galaxies and all the other stuff that astronomers can't even observe are all part of the universe. It is, simply, everything.

The star-forming nebula W51 is one of the largest "star factories" in the Milky Way galaxy. "Star factories" like this one can operate for millions of years. The cavernous red region on the right side of W51 is older, evident in the way it has already been carved out by winds from generations of massive stars (those at least 10 times the mass of our Sun). The dust and gas in the region are swept around even more when those stars die and explode as supernovas. On the nebula's younger left side, many stars are just beginning to clear away the gas and dust.

Though the universe may seem a strange place, it is not a distant one. Wherever you are right now, outer space is only 62 miles (100 kilometres) away. Day or night, whether you're indoors or outdoors, asleep, eating lunch or dozing off in class, outer space is just a few dozen miles above your head. It's below you too. About 8,000 miles (12,800 kilometres) below your feet — on the opposite side of Earth — lurks the unforgiving vacuum and radiation of outer space.

In fact, you're technically in space right now. Humans say "out in space" as if it's there and we're here, as if Earth is separate from the rest of the universe. But Earth is a planet, and it is in space and part of the universe just like the other planets. It just so happens that things live here and the environment near the surface of this particular planet is hospitable for life as we know it. Earth is a tiny, fragile exception in the cosmos. For humans and the other things living on our planet, practically the entire cosmos is a hostile and merciless environment.

This true-color image shows North and South America as they would appear from space 22,000 miles (35,000 km) above the Earth. The image is a combination of data from two satellites. The Moderate Resolution Imaging Spectroradiometer (MODIS) instrument aboard NASA's Terra satellite collected the land surface data over 16 days, while NOAA's Geostationary Operational Environmental Satellite (GOES) produced a snapshot of the Earth's clouds and the Moon.

How Old is Earth?

Our planet, Earth, is an oasis not only in space, but in time. It may feel permanent, but the entire planet is a fleeting thing in the lifespan of the universe. For nearly two-thirds of the time since the universe began, Earth did not even exist. Nor will it last forever in its current state. Several billion years from now, the Sun will expand, swallowing Mercury and Venus, and filling Earth's sky. It might even expand large enough to swallow Earth itself. It's difficult to be certain. After all, humans have only just begun deciphering the cosmos.

While the distant future is difficult to accurately predict, the distant past is slightly less so. By studying the radioactive decay of isotopes on Earth and in asteroids, scientists have learned that our planet and the solar system formed around 4.6 billion years ago.

What is the Universe Made of?

The universe contains all the energy and matter there is. Much of the observable matter in the universe takes the form of individual atoms of hydrogen, which is the simplest atomic element, made of only a proton and an electron (if the atom also contains a neutron, it is instead called deuterium). Two or more atoms sharing electrons is a molecule. Many trillions of atoms together is a dust particle. Smoosh a few tons of carbon, silica, oxygen, ice, and some metals together and you have an asteroid. Or collect 333,000 Earth masses of hydrogen and helium together, and you have a Sun-like star.

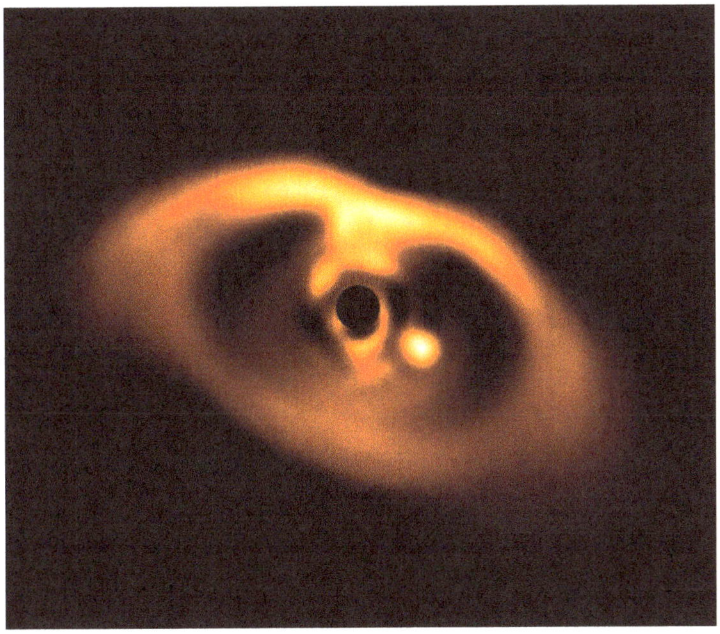

This spectacular image from the SPHERE instrument on ESO's Very Large Telescope is the first clear image of a planet caught in the very act of formation around the dwarf star PDS 70. The planet stands clearly out, visible as a bright point to the right of the centre of the image, which is blacked out by the coronagraph mask used to block the blinding light of the central star.

For the sake of practicality, humans categorize clumps of matter based on their attributes. Galaxies, star clusters, planets, dwarf planets, rogue planets, moons, rings, ringlets, comets, meteorites, raccoons — they're all collections of matter exhibiting characteristics different from one another but obeying the same natural laws.

Scientists have begun tallying those clumps of matter and the resulting numbers are pretty wild. Our home galaxy, the Milky Way, contains at least 100 billion stars, and the observable universe contains at least 100 billion galaxies. If galaxies were all the same size, that would give us 10 thousand billion billion (or 10 sextillion) stars in the observable universe.

But the universe also seems to contain a bunch of matter and energy that we can't see or directly observe. All the stars, planets, comets, sea otters, black holes and dung beetles together represent less than 5 per cent of the stuff in the universe. About 27 per cent of the remainder is dark matter, and 68 per cent is dark energy, neither of which are even remotely understood. The universe as we understand it wouldn't work if dark matter and dark energy didn't exist, and they're labelled "dark" because scientists can't seem to directly observe them. At least not yet.

Two views from Hubble of the massive galaxy cluster Cl 0024+17 (ZwCl 0024+1652) are shown. To the left is the view in visible-light with odd-looking blue arcs appearing among the yellowish galaxies. These are the magnified and distorted images of galaxies located far behind the cluster. Their light is bent and amplified by the immense gravity of the cluster in a process called gravitational lensing. To the right, blue shading has been added to indicate the location of invisible material called dark matter that is mathematically required to account for the nature and placement of the gravitationally lensed galaxies that are seen.

How our View of the Universe Changed over Time?

Human understanding of what the universe is, how it works and how vast it is has changed over the ages. For countless lifetimes, humans had little or no means of understanding the universe. Our distant ancestors instead relied upon myth to explain the origins of everything. Because our ancestors themselves invented them, the myths reflect human concerns, hopes, aspirations or fears rather than the nature of reality.

Several centuries ago, however, humans began to apply mathematics, writing and new investigative principles to the search for knowledge. Those principles were refined over time, as were scientific tools, eventually revealing hints about the nature of the universe. Only a few hundred years ago, when people began systematically investigating the nature of things, the word "scientist" didn't even exist (researchers were instead called "natural philosophers" for a time). Since then, our knowledge of the universe has repeatedly leapt forward. It was only about a century ago that astronomers first observed galaxies beyond our own, and only a half-century has passed since humans first began sending spacecraft to other worlds.

In the span of a single human lifetime, space probes have voyaged to the outer solar system and sent back the first up-close images of the four giant outermost planets and their countless moons; rovers wheeled along the surface on Mars for the first time; humans constructed a permanently crewed, Earth-orbiting space station; and the first large space telescopes delivered jaw-dropping views of more distant parts of the cosmos than ever before. In the early 21st century alone, astronomers discovered thousands of planets around other stars, detected gravitational waves for the first time and produced the first image of a black hole.

With ever-advancing technology and knowledge, and no shortage of imagination, humans continue to lay bare the secrets of the cosmos. New insights and inspired notions aid in this pursuit, and also spring from it. We have yet to send a space probe to even the nearest of the billions upon billions of other stars in the galaxy. Humans haven't even explored all the worlds in our own solar system. In short, most of the universe that can be known remains unknown.

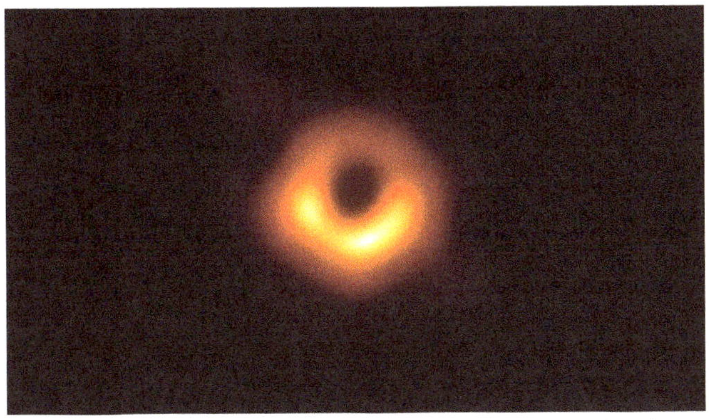

Using the Event Horizon Telescope, scientists obtained an image of the black hole at the centre of galaxy M87.

The universe is nearly 14 billion years old, our solar system is 4.6 billion years old, life on Earth has existed for maybe 3.8 billion years, and humans have been around for only a few hundred thousand years. In other words, the universe has existed roughly 56,000 times longer than our species has. By that measure, almost everything that's ever happened did so before humans existed.

Chronology

It all starts at the Big Bang, which "is a moment in time, not a point in space,". Specifically, it's the moment when time itself began, the instant from which all subsequent instants have been counted. Despite its well-known moniker, the Big Bang wasn't really an explosion but rather a period when the universe was extremely hot and dense and space began to expand outward in all directions at once. Though the model of the Big Bang states that the universe was an infinitely small point of infinite density, that's just a hand-wavy way of saying that we don't quite know what was going on then. Mathematical infinities don't make sense in physics equations, so the Big Bang is really the point at which our current understanding of the universe breaks down.

Cosmic Inflation Era

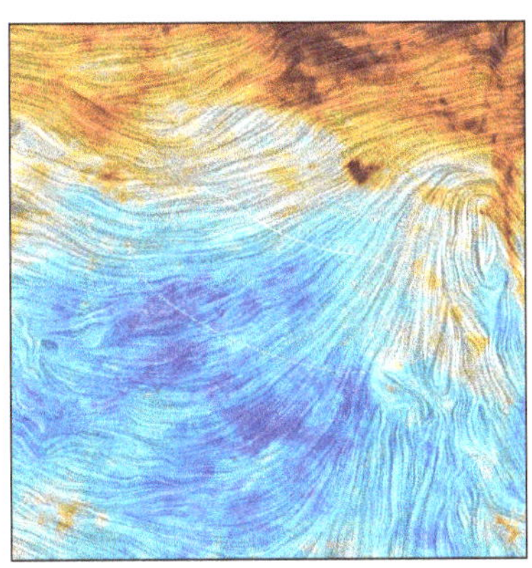

The universe's next trick was to grow really big really fast. Within the first 0.000000000000000000000000000001 (that's a decimal point with 30 zeros before the 1) seconds after the Big

Bang, the cosmos could have expanded exponentially in size, driving apart areas of the universe that had previously been in close contact. This era, known as inflation, remains hypothetical, but cosmologists like the idea because it explains why far-flung regions of space appear so similar to one another, despite being separated by vast distances. Back in 2014, a team thought they had found a signal of this expansion in light from the early universe. But the results later turned out to be something much more mundane: interfering interstellar dust.

Quark-Gluon Plasma

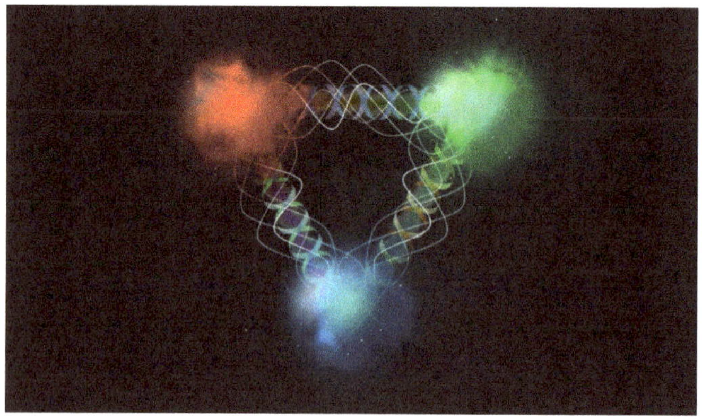

A few milliseconds after the beginning of time, the early universe was really hot — we're talking between 7 trillion and 10 trillion degrees Fahrenheit (4 trillion and 6 trillion degrees Celsius) hot. At such temperatures, elementary particles called quarks, which are normally bound tightly inside of protons and neutrons, wandered around freely. Gluons, which carry a fundamental force known as the strong force, were mixed in with these quarks in a soupy primordial fluid that permeated the cosmos. Researchers have managed to create similar conditions in particle accelerators on Earth. But the difficult-to-achieve state only ever lasted a few fractions of a second, in terrestrial atom smashers as well as in the early universe.

The Early Epoch

There was a lot of action in the next stage of time, which began around a few thousandths of a second after the Big Bang. As the cosmos expanded, it cooled, and soon conditions were clement enough for quarks to come together into protons and neutrons. One second after the Big Bang, the universe's density dropped enough that neutrinos — the lightest and least-interacting fundamental particle — could fly forward without hitting anything, creating what's known as the cosmic neutrino background, which scientists have yet to detect.

The First Atoms

For the first 3 minutes of the universe's life, protons and neutrons fused together, forming an isotope of hydrogen called deuterium as well as helium and a tiny amount of the next-lightest element, lithium. But once the temperature fell, this process stopped. Finally, 380,000 years after the Big Bang, things were cool enough so that hydrogen and helium could combine with free electrons, creating the first neutral atoms. Photons, which had previously run into the electrons, could now move without interference, creating the cosmic microwave background (CMB), a relic from this era that was first detected in 1965.

The Dark Ages

For a very long time, nothing in the universe gave off light. This period, which lasted around 100 million years, is known as the Cosmic Dark Ages. This epoch remains extremely difficult to study because astronomers' knowledge of the universe comes almost entirely from starlight. Without any stars, it's difficult to know what went on.

The First Stars

By around 180 million years after the Big Bang, hydrogen and helium began to collapse into large spheres, generating infernal temperatures in their cores that lit up into the first stars. The universe entered a period known as Cosmic Dawn, or re-ionization, because the hot photons radiated by early stars and galaxies broke neutral hydrogen atoms in interstellar space into protons and electrons, a process known as ionization. Just how long re-ionization lasted is difficult to say. Because it occurred so early, its signals are obscured by later gas and dust, so the best scientists can say is that it was over by around 500 million years after the Big Bang.

Large-Scale Structure

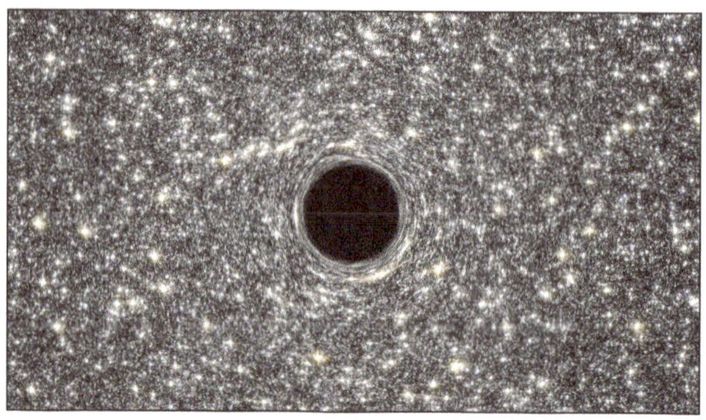

Here's where the universe gets down to business, or at least the familiar business we know about today. Small early galaxies began to merge together into larger galaxies and, around 1 billion years after the Big Bang, supermassive black holes formed in their centres. Bright quasars, which produce intense beacons of light that can be seen from 12 billion light-years away, turned on.

The Universe's Middle Years

The universe continued to evolve over the next several billion years. Spots of higher density from the primordial universe gravitationally attracted matter to themselves. These slowly grew into galactic clusters and long strands of gas and dust, producing a beautiful filamentary cosmic web that can be seen today.

Birth of the Solar System

About 4.5 billion years ago, in one particular galaxy, a cloud of gas collapsed down into yellow star with a system of rings around it. These rings coalesced into eight planets, plus various comets, asteroids, dwarf planets, and moons, forming a familiar stellar system. The planet third from the central star managed to either retain a ton of water after this process, or else comets later delivered a deluge of ice and water.

Age

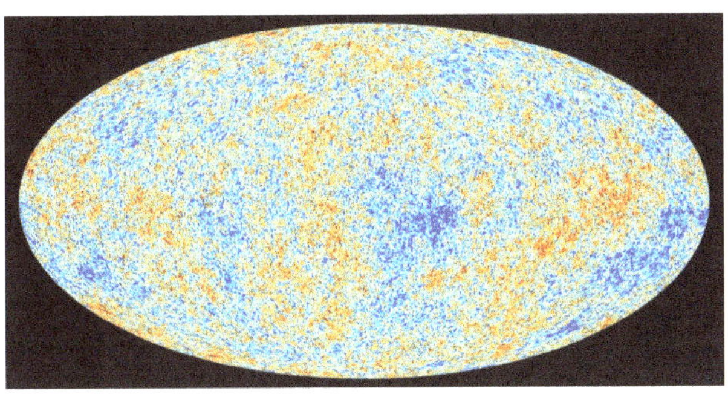

A 2013 map of the background radiation left over from the Big Bang, taken by the ESA's Planck spacecraft, captured the oldest light in the universe. This information helps astronomers determine the age of the universe.

Age may only be a number, but when it comes to the age of the universe, it's a pretty important one. According to research, the universe is approximately 13.8 billion years old. How did scientists determine how many candles to put on the universe's birthday cake? They can determine the age of the universe using two different methods: by studying the oldest objects within the universe and measuring how fast it is expanding.

Age Limits

The universe cannot be younger than the objects contained inside of it. By determining the ages of the oldest stars, scientists are able to put a limit on the age.

The life cycle of a star is based on its mass. Massive stars burn faster than their lower-mass siblings. A star 10 times as massive as the sun will burn through its fuel supply in 20 million years, while a star with half the sun's mass will last more than 20 billion years. The mass also affects the brightness, or luminosity, of a star; more massive stars are brighter.

Known as Population III stars, the first stars were massive and short-lived. They contained only hydrogen and helium, but through fusion began to create the elements that would help to build the next generation of stars. Scientists have been hunting for traces of the first stars for decades.

> "Those stars were the ones that formed the first heavy atoms that ultimately allowed us to be here."

"The detection of dust in the early universe provides new information on when the first supernovae exploded and hence the time when the first hot stars bathed the universe in light,". "Determining the timing of this 'cosmic dawn' is one of the holy grails of modern astronomy, and it can be indirectly probed through the study of early interstellar dust."

Early stars aren't the only way to place limits on the age of the universe. Dense collections of stars known as globular clusters have similar characteristics. The oldest known globular clusters have stars with ages that appear to be between 11 and 14 billion years old. The wide range comes from problems in pinpointing the distances to the clusters, which affects estimates of brightness and thus mass. If the cluster is farther away than scientists have measured, the stars would be brighter, thus more massive, and thus younger than calculated.

"Just like archaeologists use fossils to reconstruct the history of the Earth, astronomers use globular clusters to reconstruct the history of the galaxy,". "There are only about 150 globular clusters known in the Milky Way Galaxy, so each of these globular clusters is an important tracer of the galactic halo and the formation of the Milky Way Galaxy." The uncertainty still creates a limit to the age of the universe; it must be at least 11 billion years old. It can be older, but not younger.

Expansion of the Universe

The universe we live in is not flat and unchanging, but constantly expanding. If the expansion rate is known, scientists can work backwards to determine the universe's age, much like police officers can unravel the initial conditions that resulted in a traffic accident. Thus, finding the expansion rate of the universe — a number known as the Hubble constant — is key.

A number of factors determine the value of this constant. The first is the type of matter that dominates the universe. Scientists must determine the proportion of regular and dark matter to dark energy. Density also plays a role. A universe with a low density of matter is older than a matter-dominated one.

To determine the density and composition of the universe, scientists rely on missions. By measuring the thermal radiation left over from the Big Bang, missions such as these are able to determine the density, composition and expansion rate of the universe. The leftover radiation is known as the cosmic microwave background, and both WMAP and Planck have mapped it.

In 2012, WMAP estimated the age of the universe to be 13.772 billion years, with an uncertainty of 59 million years. In 2013, Planck measured the age of the universe at 13.82 billion years. Both of these fall within the lower limit of 11 billion years independently derived from the globular clusters, and both have smaller uncertainties than that number.

For thousands of years, astronomers wrestled with basic questions about the size and age of the universe. Does the universe go on forever, or does it have an edge somewhere? Has it always existed, or did it come to being sometime in the past? In 1929, Edwin Hubble, an astronomer at Caltech, made a critical discovery that soon led to scientific answers for these questions: he discovered that the universe is expanding.

The ancient Greeks recognized that it was difficult to imagine what an infinite universe might look like. But they also wondered that if the universe was finite, and you stuck out your hand at the edge, where would your hand go? The Greeks' two problems with the universe represented a paradox - the universe had to be either finite or infinite, and both alternatives presented problems.

After the rise of modern astronomy, another paradox began to puzzle astronomers. In the early 1800s, German astronomer Heinrich Olbers argued that the universe must be finite. If the Universe were infinite and contained stars throughout, Olbers said, and then if you looked in any particular direction, your line-of-sight would eventually fall on the surface of a star. Although the apparent size of a star in the sky becomes smaller as the distance to the star increases, the brightness of this smaller surface remains a constant. Therefore, if the Universe were infinite, the whole surface of the night sky should be as bright as a star. Obviously, there are dark areas in the sky, so the universe must be finite.

But, when Isaac Newton discovered the law of gravity, he realized that gravity is always attractive. Every object in the universe attracts every other object. If the universe truly were finite, the attractive forces of all the objects in the universe should have caused the entire universe to collapse on itself. This clearly had not happened, and so astronomers were presented with a paradox.

When Einstein developed his theory of gravity in the General Theory of Relativity, he thought he ran into the same problem that Newton did: his equations said that the universe should be either expanding or collapsing, yet he assumed that the universe was static. His original solution contained a constant term, called the cosmological constant, which cancelled the effects of gravity on very large scales, and led to a static universe. After Hubble discovered that the universe was expanding, Einstein called the cosmological constant his "greatest blunder."

At around the same time, larger telescopes were being built that were able to accurately measure the spectra, or the intensity of light as a function of wavelength, of faint objects. Using these new data, astronomers tried to understand the plethora of faint, nebulous objects they were observing. Between 1912 and 1922, astronomer Vesto Slipher at the Lowell Observatory in Arizona discovered that the spectra of light from many of these objects were systematically shifted to longer wavelengths, or redshifted. A short time later, other astronomers showed that these nebulous objects were distant galaxies.

The Discovery of the Expanding Universe

Meanwhile, other physicists and mathematicians working on Einstein's theory of gravity discovered the equations had some solutions that described an expanding universe. In these solutions, the light coming from distant objects would be redshifted as it travelled through the expanding universe. The redshift would increase with increasing distance to the object.

Edwin Hubble

In 1929 Edwin Hubble, working at the Carnegie Observatories in Pasadena, California, measured the redshifts of a number of distant galaxies. He also measured their relative distances by measuring the apparent brightness of a class of variable stars called Cepheids in each galaxy. When he plotted redshift against relative distance, he found that the redshift of distant galaxies increased as a linear function of their distance. The only explanation for this observation is that the universe was expanding.

Once scientists understood that the universe was expanding, they immediately realized that it would have been smaller in the past. At some point in the past, the entire universe would have been a single point. This point, later called the big bang, was the beginning of the universe as we understand it today.

The expanding universe is finite in both time and space. The reason that the universe did not collapse, as Newton's and Einstein's equations said it might, is that it had been expanding from the moment of its creation. The universe is in a constant state of change. The expanding universe, a new idea based on modern physics, laid to rest the paradoxes that troubled astronomers from ancient times until the early 20th Century.

Properties of the Expanding Universe

The equations of the expanding universe have three possible solutions, each of which predicts a different eventual fate for the universe as a whole. Which fate will ultimately befall the universe can be determined by measuring how fast the universe expands relative to how much matter the universe contains.

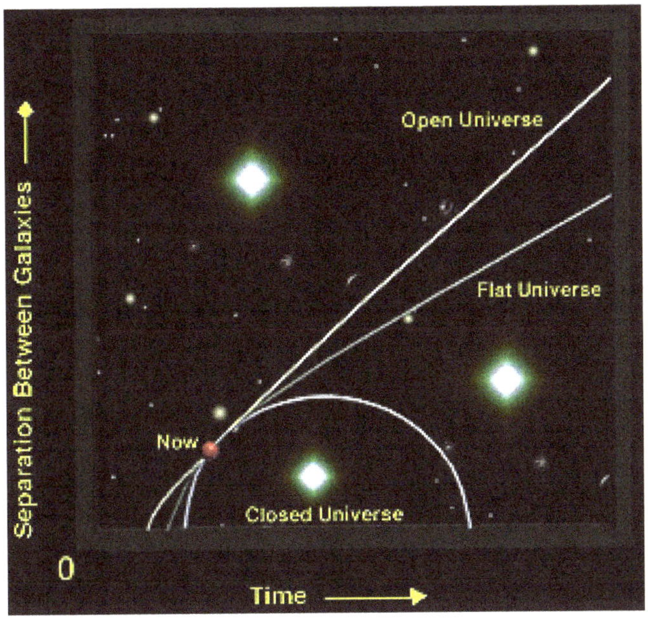

The three possible types of expanding universes are called open, flat, and closed universes. If the universe were open, it would expand forever. If the universe were flat, it would also expand forever, but the expansion rate would slow to zero after an infinite amount of time. If the universe were closed, it would eventually stop expanding and re-collapse on itself, possibly leading to another big bang. In all three cases, the expansion slows, and the force that causes the slowing is gravity.

A simple analogy to understand these three types of universes is to consider a spaceship launched from the surface of the Earth. If the spaceship does not have enough speed to escape the Earth's gravity, it will eventually fall back to Earth. This is analogous with a closed universe that re-collapses. If the spaceship is given enough speed so that it has just enough energy to escape, then at an infinite distance away from the Earth, it will come to a stop (this is the flat universe). And lastly, if the ship is launched with more than enough energy to escape, it will always have some speed, even when it is an infinite distance away (the open universe).

The Fate of the Universe

For the last eighty years, astronomers have been making increasingly accurate measurements of two important cosmological parameters: H_0 - the rate at which the universe expands - and w - the average density of matter in the universe. Knowledge of both of these parameters will tell which of the three models describes the universe we live in, and thus the ultimate fate of our universe. The Sloan Digital Sky Survey, with its large systematic measurement of the galaxy density in the Universe, should enable astronomers to precisely measure the density parameter ω.

The Heavy Elements

Astronomers are not only interested in the fate of the universe; they are also interested in understanding its present physical state. One question they try to answer is why the universe is primarily composed of hydrogen and helium, and what is responsible for the relatively small concentration of the heavier elements.

With the rise of nuclear physics in the 1930s and 40s, scientists started to try to explain the abundances of heavier elements by assuming they were synthesized out of primordial hydrogen in the early universe. In the late 1940s, American physicists George Gamow, Robert Herman, and Ralph Alpher realized that long ago, the universe was much hotter and denser. They made calculations to show whether nuclear reactions that took place at those higher temperatures could have created the heavy elements.

Unfortunately, with the exception of helium, they found that it was impossible to form heavier elements in any appreciable quantity. Today, we understand that heavy elements were synthesized either in the cores of stars or during supernovae, when a large dying star implodes.

Gamow, Herman, and Alpher did realize, though, that if the universe were hotter and denser in the past, radiation should still be left over from the early universe. This radiation would have a well-defined spectrum (called a blackbody spectrum) that depends on its temperature. As the universe expanded, the spectrum of this light would have been redshifted to longer wavelengths, and the temperature associated with the spectrum would have decreased by a factor of over one thousand as the universe cooled.

The Cosmic Microwave Background Radiation

In 1963, Arno Penzias and Robert Wilson, two scientists in Holmdale, New Jersey, were working on a satellite designed to measure microwaves. When they tested the satellite's antenna, they found mysterious microwaves coming equally from all directions. At first, they thought something was wrong with the antenna. But after checking and rechecking, they realized that they had discovered something real. What they discovered was the radiation predicted years earlier by Gamow, Herman, and Alpher. The radiation that Penzias and Wilson discovered called the Cosmic Microwave Background Radiation, convinced most astronomers that the Big Bang theory was correct. For discovering the Cosmic Microwave Background Radiation, Penzias and Wilson were awarded the 1978 Nobel Prize in Physics.

After Penzias and Wilson found the Cosmic Microwave Background Radiation, astrophysicists began to study whether they could use its properties to study what the universe was like long ago. According to Big Bang theory, the radiation contained information on how matter was distributed over ten billion years ago, when the universe was only 500,000 years old.

At that time, stars and galaxies had not yet formed. The Universe consisted of a hot soup of electrons and atomic nuclei. These particles constantly collided with the photons that made up the background radiation, which then had a temperature of over 3000 C.

Soon after, the Universe expanded enough, and thus the background radiation cooled enough, so that the electrons could combine with the nuclei to form atoms. Because atoms were electrically neutral, the photons of the background radiation no longer collided with them.

When the first atoms formed, the universe had slight variations in density, which grew into the density variations we see today - galaxies and clusters. These density variations should have led to slight variations in the temperature of the background radiation, and these variations should still be detectable today. Scientists realized that they had an exciting possibility: by measuring the temperature variations of the Cosmic Microwave Background Radiation over different regions of the sky, they would have a direct measurement of the density variations in the early universe, over 10 billion years ago.

Density Variations in the Early Universe

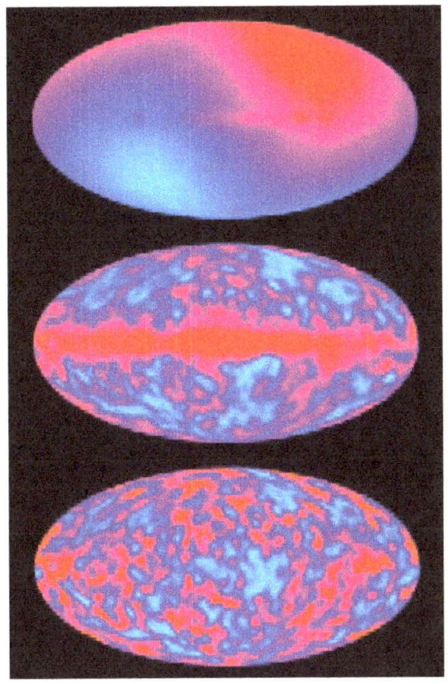

Map of the sky as observed by COBE. The map on the bottom shows the temperature variations of the background radiation.

In 1990, a satellite called the Cosmic Microwave Background Explorer (COBE) measured background radiation temperatures over the whole sky. COBE found variations that amounted to only about 5 parts in 100,000, but revealed the density fluctuations in the early universe.

The initial density variations would be the seeds of structure that would grow over time to become the galaxies, clusters of galaxies, and super clusters of galaxies observed today by the Sloan Digital Sky Survey. With the Sloan data, along with data from COBE, astronomers will be able to reconstruct the evolution of structure in the universe over the last 10 to 15 billion years. With this information, we will have a deep understanding of the history of the universe, which will be an almost unbelievable scientific and intellectual achievement.

But measuring the evolution of the density variations in the universe still does not answer the most important question: why does the universe contain these differences in density in the first place? To answer this question, astronomers and astrophysicists must understand the nature of the density variations and construct theories of the origin of the universe that predict how these variations should occur.

Cosmic Microwave Background

Cosmic microwave background (CMB), also called cosmic background radiation is the electromagnetic radiation filling the universe that is a residual effect of the big bang 13.8 billion years ago. Because the expanding universe has cooled since this primordial explosion, the background radiation is in the microwave region of the electromagnetic spectrum.

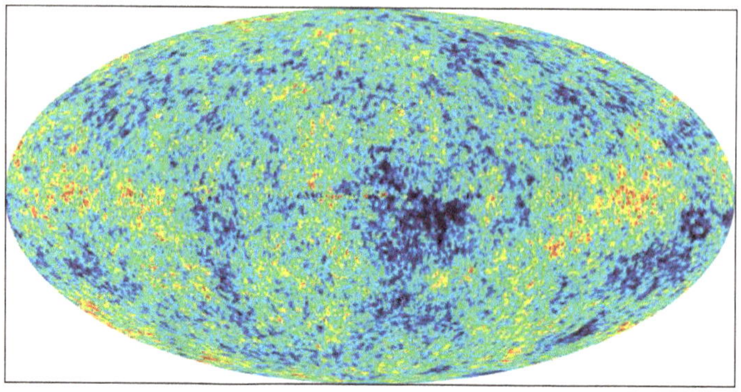

Wilkinson Microwave Anisotropy Probe.

A full-sky map produced by the Wilkinson Microwave Anisotropy Probe (WMAP) showing cosmic background radiation, a very uniform glow of microwaves emitted by the infant universe more than 13 billion years ago. Colour differences indicate tiny fluctuations in the intensity of the radiation, a result of tiny variations in the density of matter in the early universe. According to inflation theory, these irregularities were the "seeds" that became the galaxies. WMAP's data support the big bang and inflation models.

Beginning in 1948, the American cosmologist George Gamow and his co-workers, Ralph Alpher and Robert Herman, investigated the idea that the chemical elements might have been synthesized by thermonuclear reactions that took place in a primeval fireball. According to their calculations, the high temperature associated with the early universe would have given rise to a thermal radiation field, which has a unique distribution of intensity with wavelength (known as Planck's radiation law), that is a function only of the temperature. As the universe expanded, the temperature would have dropped, each photon being redshifted by the cosmological expansion to longer wavelength, as the American physicist Richard C. Tolman had already shown in 1934. By the present epoch the radiation temperature would have dropped to very low values, about 5 kelvins above absolute zero (0 Kelvin [K], or −273 °C [−460 °F]) according to the estimates of Alpher and Herman.

Interest in these calculations waned among most astronomers when it became apparent that the lion's share of the synthesis of elements heavier than helium must have occurred inside stars rather than in a hot big bang. In the early 1960s physicists at Princeton University, New Jersey, as well as in the Soviet Union, took up the problem again and began to build a microwave receiver that might detect, in the words of the Belgian cleric and cosmologist Georges Lemaître, "the vanished brilliance of the origin of the worlds."

The actual discovery of the relict radiation from the primeval fireball, however, occurred by accident. In experiments conducted in connection with the first Telstar communication satellite, two scientists, Arno Penzias and Robert Wilson, of the Bell Telephone Laboratories, Holmdel, New Jersey, measured excess radio noise that seemed to come from the sky in a completely isotropic fashion (that is, the radio noise was the same in every direction). When they consulted Bernard Burke of the Massachusetts Institute of Technology, Cambridge, about the problem, Burke realized that Penzias and Wilson had most likely found the cosmic background radiation that Robert H. Dicke, P.J.E. Peebles, and their colleagues at Princeton were planning to search for. Put in touch with one another, the two groups published simultaneously in 1965 papers detailing the prediction and discovery of a universal thermal radiation field with a temperature of about 3 K.

Precise measurements made by the Cosmic Background Explorer (COBE) satellite launched in 1989 determined the spectrum to be exactly characteristic of a blackbody at 2.735 K. The velocity of the satellite about Earth, Earth about the Sun, the Sun about the Galaxy, and the Galaxy through the universe actually makes the temperature seem slightly hotter (by about one part in 1,000) in the direction of motion rather than away from it. The magnitude of this effect—the so-called dipole anisotropy—allows astronomers to determine that the Local Group (the group of galaxies containing the Milky Way Galaxy) is moving at a speed of about 600 km per second (km/s; 400 miles per second [miles/s]) in a direction that is 45° from the direction of the Virgo cluster of galaxies. Such motion is not measured relative to the galaxies themselves (the Virgo galaxies have an average velocity of recession of about 1,000 km/s [600 miles/s] with respect to the Milky Way system) but relative to a local frame of reference in which the cosmic microwave background radiation would appear as a perfect Planck spectrum with a single radiation temperature.

The COBE satellite carried instrumentation aboard that allowed it to measure small fluctuations in intensity of the background radiation that would be the beginning of structure (i.e., galaxies and clusters of galaxies) in the universe. The satellite transmitted an intensity pattern in angular projection at a wavelength of 0.57 cm after the subtraction of a uniform background at a temperature of 2.735 K. Bright regions at the upper right and dark regions at the lower left showed the dipole asymmetry. A bright strip across the middle represented excess thermal emission from the Milky Way. To obtain the fluctuations on smaller angular scales, it was necessary to subtract both the dipole and the galactic contributions. An image was obtained showing the final product after the subtraction. Patches of light and dark represented temperature fluctuations that amount to about one part in 100,000—not much higher than the accuracy of the measurements. Nevertheless, the statistics of the distribution of angular fluctuations appeared different from random noise, and so the members of the COBE investigative team found the first evidence for the departure from exact isotropy that theoretical cosmologists long predicted must be there in order for galaxies and clusters of galaxies to condense from an otherwise structure less universe. These fluctuations correspond to distance scales on the order of 10^9 light-years across (still larger than the largest material structures seen in the universe, such as the enormous grouping of galaxies dubbed the "Great Wall").

The Wilkinson Microwave Anisotropy Probe (WMAP) was launched in 2001 to observe the fluctuations seen by COBE in greater detail and with more sensitivity. The conditions at the beginning of the universe left their imprint on the size of the fluctuations. WMAP's accurate measurements showed that the early universe was 63 per cent dark matter, 15 per cent photons, 12 per cent atoms, and 10 per cent neutrinos. Today the universe is 72.6 per cent dark energy, 22.8 per cent dark matter, and 4.6 per cent atoms. Although neutrinos are now a negligible component of the universe, they form their own cosmic background, which was discovered by WMAP. WMAP also showed that the first stars in the universe formed half a billion years after the big bang.

Isotropy in the Cosmic Background

Apart from the small fluctuations (one part in 100,000), the observed cosmic microwave background radiation exhibits a high degree of isotropy, a 0th order fact that presents both satisfaction and difficulty for a comprehensive theory. On the one hand, it provides a strong justification for the assumption of homogeneity and isotropy that is common to most cosmological models. On the other hand, such homogeneity and isotropy are difficult to explain because of the "light-horizon" problem. In the context of the cosmic microwave background, the problem can be expressed as follows. Consider the background radiation coming to an observer from any two opposite sides of the sky. Clearly, whatever are the ultimate sources (hot plasma) of this radiation, the photons, traveling at the speed of light since their emission by the plasma, have only had time to reach Earth now? The matter on one side of the sky could not have had time to have "communicated" with the matter on the other side (they are beyond each other's light horizon), so how is it possible (with respect to an observer in the right rest frame) that they "know" to have the same temperature to a precision approaching one part in 100,000? What accounts for the high degree of angular isotropy of the cosmic microwave background?

A mechanism called "inflation" offers an attractive way out of this dilemma. The basic idea is that at high energies, matter is better described by fields than by classical means. The contribution of a field to the energy density (and therefore the mass density) and the pressure of the vacuum state need not have been zero in the past, even if it is today. During the time of super unification (Planck era, 10^{-43} second) or grand unification (GUT era, 10^{-35} second), the lowest-energy state for this field may have corresponded to a "false vacuum," with a combination of mass density and negative pressure that results gravitationally in a large repulsive force. In the context of Einstein's theory of general relativity, the false vacuum may be thought of alternatively as contributing a cosmological constant about 10100 times larger than it can possibly be today. The corresponding repulsive force causes the universe to inflate exponentially, doubling its size roughly once every 10^{-43} or 10^{-35} second. After at least 85 doublings, the temperature, which started out at 10^{32} or 10^{28} K, would have dropped to very low values near absolute zero.

Inflation provides a mechanism for understanding the overall isotropy of the cosmic microwave background because the matter and radiation of the entire observable universe were in good thermal contact (within the cosmic event horizon) before inflation and therefore acquired the same thermodynamic characteristics. Rapid inflation carried different portions outside their individual event horizons. When inflation ended and the universe reheated and resumed normal expansion, these different portions, through the natural passage of time, reappeared on our horizon. Through the observed isotropy of the cosmic microwave background, they are inferred still to have the same temperatures.

Given the measured radiation temperature of 2.735 K, the energy density of the cosmic microwave background can be shown to be about 1,000 times smaller than the average rest-energy density of ordinary matter in the universe. Thus, the current universe is matter-dominated. If one goes back in time to redshift z, the average number densities of particles and photons were both bigger by the same factor $(1 + z)^3$ because the universe was more compressed by this factor, and the ratio of these two numbers would have maintained its current value of about one hydrogen nucleus, or proton, for every 10^9 photons. The wavelength of each photon, however, was shorter by the factor $1 + z$ in the past than it is now; therefore, the energy density of radiation increases faster by one factor of $1 + z$ than the rest-energy density of matter. Thus, the radiation energy density becomes comparable to the energy density of ordinary matter at a redshift of about 1,000. At redshifts larger than 10,000, radiation would have dominated even over the dark matter of the universe. Between these two values, at a redshift of about 1,090, radiation would have decoupled from matter when hydrogen recombined. It is not possible to use photons to observe redshifts larger than this, because the cosmic plasma at temperatures above 4,000 K is essentially opaque before recombination. These photons from a redshift of 1,090 form the cosmic microwave background.

Cosmic Background Radiation

The universe is filled with blackbody radiation whose temperature now is $T_0 = 2:725 \pm 0:002$ K, so the frequency of the peak brightness is $\nu_{max} \approx 160\,GHz$. This cosmic microwave background (CMB) is a relic of the "big bang" creation of the universe and reveals precise values for a host of cosmological parameters. Before the CMB was first observed, George Gamow and his students calculated the hydrogen and helium abundances produced in the presence of blackbody radiation when the universe was very hot (Alpher, R., Bethe, H., & Gamow, G. 1948, Phys. Rev. 73, 803), and Alpher, R. A., & Herman, R. 1948, Nature, 162, 774 contains the first estimate $T_0 \approx 5$ K for the present temperature of the universe.

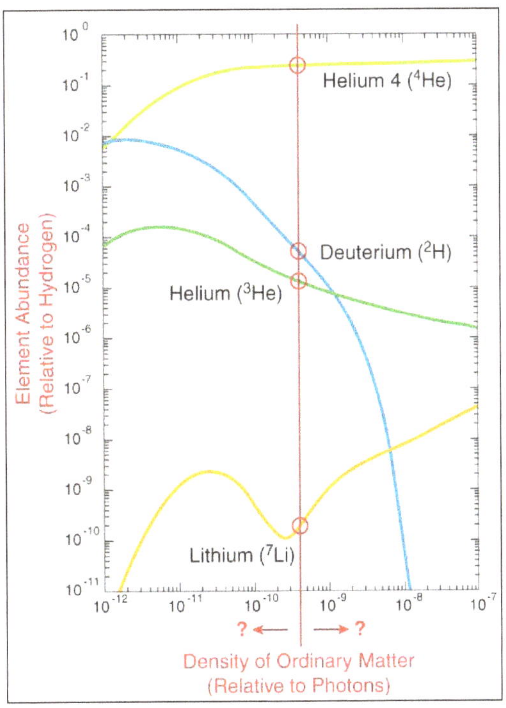

The energy density of the early universe was dominated by hot blackbody radiation. By $t \approx 1$ s the temperature fell to $T \approx 10^{10}$ K and neutrons decayed into protons and electrons or combined with protons to make deuterium nuclei, most of which combined to make helium nuclei during the first three minutes. The chemical composition of the early universe is a function of the photon/baryon ratio. Baryonic matter now accounts for about 4.6% of the energy density of the universe.

Although 5K is a very strong signal by radio astronomy standards, the CMB is not easy to detect because it is very nearly isotropic. Unlike a compact radio source, the CMB cannot be detected by a differential measurement on two nearby regions of sky; it is difficult to distinguish from other noise sources in the receiver and the radio telescope. It was accidentally discovered by Arno Penzias and Robert Wilson at Bell Labs. They reported careful measurements indicating an isotropic "excess antenna temperature" produced by their horn antenna at 4080 MHz Subsequent measurements, especially by the COBE (COsmic Background Explorer) satellite, have demonstrated that this "excess noise" matches a blackbody spectrum within 50 parts per million well into the far infrared. The largest (but still very small) anisotropies are (1) a dipole anisotropy caused by the Earth's motion relative to the CMB and (2) contamination by sources in the galactic plane.

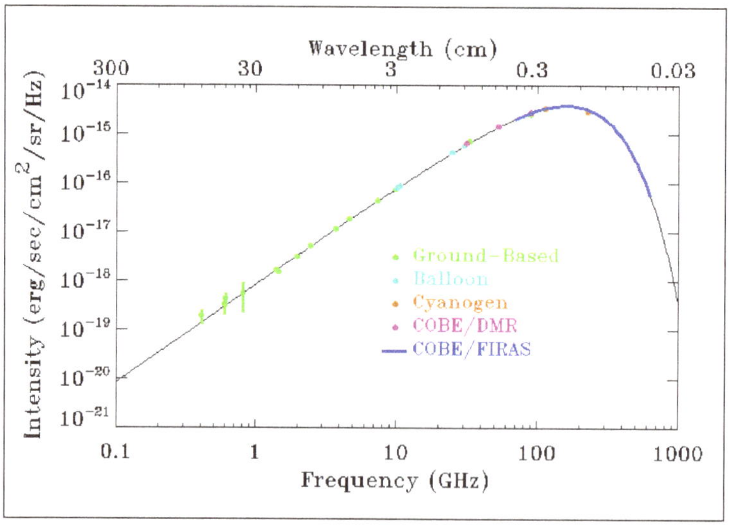

The observed CMB spectrum is a very close match to a T = 2:725 K blackbody.

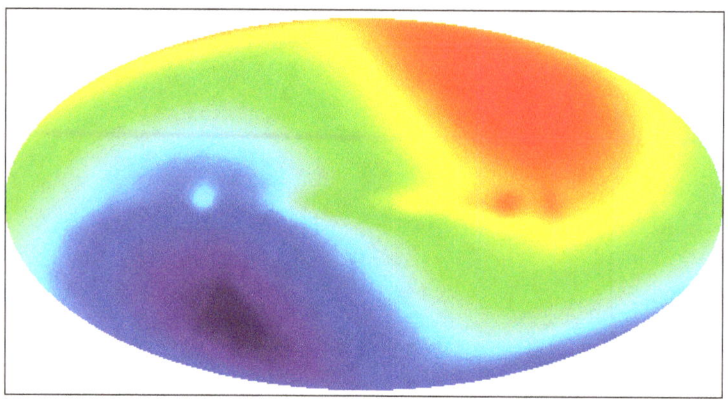

The CMB is nearly isotropic. This COBE all-sky image in galactic coordinates spans only the temperature range 2 : 275 ± 0 : 004 K. The dipole anisotropy is caused by the Earth's 600 km s^{-1} motion, and some foreground emission in the galactic plane is visible.

How does the blackbody CMB evolve as the universe expands? The universe is spatially homogenous and isotropic, so the properties of the CMB can be calculated by considering what happens to blackbody radiation in a small cube whose side length a (t) slowly grows with time t since the big bang. "Small" here means only much smaller than the Hubble distance $c / H_0 3.00\times10^5\, km\, s^{-1} / 73\, km\, s^{-1}\, Mpc^{-1} \approx 4.1\times10^3\, Mpc$ so that relativistic effects can be ignored. Although our imaginary cube has no walls, CMB radiation escaping from the cube is balanced by CMB radiation entering the cube from similar adjacent cubes. Consequently the energy density and spectrum of CMB radiation within the cube is exactly the same as that in an adiabatically expanding cubical cavity having opaque walls. The radiation remains in equilibrium during the slow expansion, so it remains blackbody radiation, but its temperature changes. Blackbody radiation in a slowly expanding cube can be described by classical thermodynamics. The first law of thermodynamics requires energy conservation as the total heat Q, internal energy U, and work W done by the radiation in the cube change as the volume V = a^3 of the cube grows by a small amount dV:

$$dQ = dU + dW = dU + pdV$$

Where, p is the pressure exerted by the radiation.

Radiation exerts pressure because photons of frequency v have momentum hv/c as well as energy hv. Consider a small flat mirror of unit area on the inside surface of the cavity. According to Kirchhoff's law, the mirror emits no radiation, so only reflections can cause radiation pressure. Each cavity photon approaching the mirror at an angle θ from the normal to the mirror will be reflected at the same angle and transfer $2hv\cos\theta$ of momentum to the mirror. The radiation pressure equals the rate of momentum transfer per unit area. Blackbody radiation is isotropic so the radiation pressure p_v produced by photons of frequency v is:

$$p_v = \frac{2}{c}\int B_v \cos^2\theta d\Omega,$$

Where the second cos θ factor comes from the projected area of the mirror, as seen by the incoming photon. The total radiation pressure is,

$$p = \int_0^\infty p_v dv = \frac{2}{c}\int_0^\infty B_v dv \int_0^{2\pi} d\phi \int_0^{\pi/2} \cos^2\theta \sin\theta\, d\theta = \frac{4\pi B}{c}\int_0^1 x^2 dx$$

Substituting the radiation energy density $u = 4\pi\ B = c$ and integrating relates the radiation pressure and energy density of isotropic radiation:

$$p = \frac{u}{3}.$$

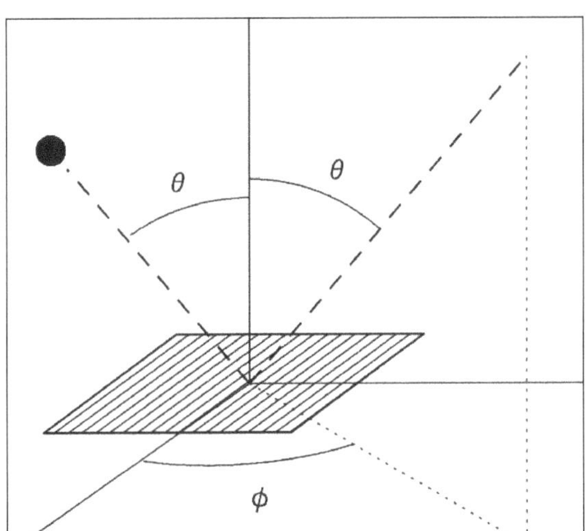

Reflection of a photon by a mirror on the cavity wall.

The second law of thermodynamics relates the change in entropy S to the change in heat Q and the temperature T:

$$dS = \frac{dQ}{T}.$$

Using the first law to eliminate dQ and converting from pressure p to energy density u gives,

$$dS = \frac{dU}{T} + p\frac{dV}{T} = \frac{d(uV)}{T} + \frac{u}{3}\frac{dV}{T} = \frac{udV}{T} + \frac{Vdu}{T} + \frac{udV}{3T}$$

$$dS = \frac{4u}{3T}dV + \frac{V}{T}\frac{du}{dT}dT$$

Setting dS = 0 for an adiabatic expansion and substituting $u = 4\sigma T^4 / c$ yields,

$$\frac{dV}{V} = -3\frac{dT}{T}.$$

Inserting the volume V = a^3 for a cube of side length a gives $dT/T = -da/a$ so we find that the CMB temperature T (t) is inversely proportional to the scale size of the universe:

$$T(t) \propto a(t)^{-1}.$$

The total number of photons in the expanding cube is unchanged, but the wavelength¸ of each photon grows linearly with the scale size a (t). Astronomers frequently use the term redshift defined by,

$$z \equiv \frac{\lambda_0 - \lambda_0}{\lambda_0} = \frac{\lambda_0}{\lambda_0} - 1.$$

Where λ_0 is wavelength emitted by a source at redshift z and λ_0 is the observed wavelength at z = 0. The redshift z and scale size a are related by:

$$(1 + z) = a^{-1}.$$

Thus the temperature of the CMB at any redshift z is:

$$T = T_0 (1 + z),$$

Where $T_0 \approx 2:725K$.

Prior to the time corresponding to the redshift $z_* = 1091 \pm 1$ the temperature was T > 3000 K, high enough to ionize the hydrogen atoms filling the universe and make the universe opaque to the CMB. This "wall" beyond which we cannot see is called the surface of last scattering. At the so-called recombination era when the age of the universe was $t_* = 379 \pm 5 \times 10^3$ yr. almost all of the free protons and electrons combined to form neutral hydrogen atoms and the universe became transparent to the CMB. The CMB photons that we see today were last scattered at that time and have been traveling in straight lines ever since, so an image of the CMB shows its temperature distribution at $z_* \approx 1091$.

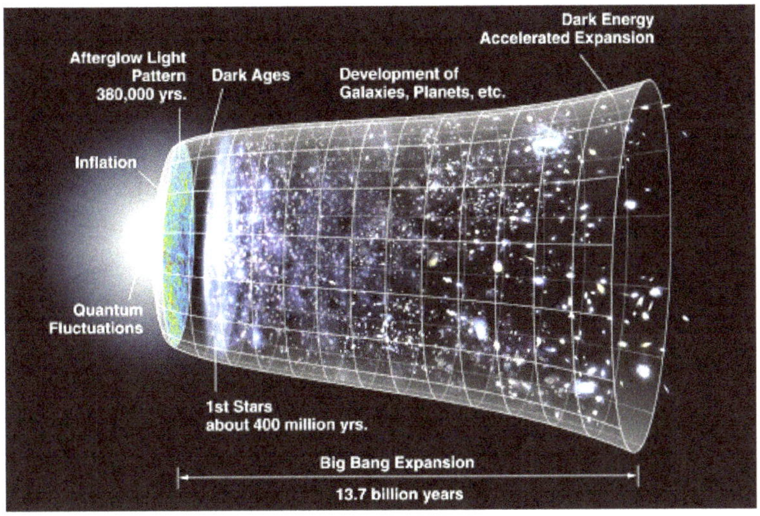

A brief history of the universe.

Intrinsic Anisotropy

The time t_* can be calculated because the physics of the homogeneous and isotropic early universe was so simple. During that time, small inhomogeneities in the density of dark matter were amplified by gravity. Baryons tended to fall into the gravitational potential wells, and the CMB was coupled to this flow by Thomson scattering off the free electrons prior to $t_?$. Radiation pressure resisted this infall, setting up baryon acoustic oscillations analogous the sound vibrations in air, but with a sound speed $c_s \approx c = \sqrt{3}$. The strongest oscillations in the CMB brightness at recombination have wavelengths:

$$\lambda_* \approx 2c_s t_* \approx \frac{2ct_*}{\sqrt{3}} \approx \frac{2*379\times10^3}{\sqrt{3}}\text{ly} \approx 438\times10^3\,\text{ly},$$

This is about 134 kpc. After recombination, the photons decoupled and this "standard ruler" grew by a factor $1+z_* \approx 1092$ to its present size of 146 Mpc. The power spectrum of CMB fluctuations observed by WMAP reveal that the angular size of this ruler is $\theta_* = 0:010388 \pm 0:000027$ rad, so the angular-size distance to the surface of last scattering is $D_A = \lambda_* = \theta_* \approx 14.1 \pm 0.16$ Gpc. The largest extragalactic distance is actually the best measured! If we live in a flat ¤CDM universe, its present age is t = 13. 75 ± 0.13 Gyr.

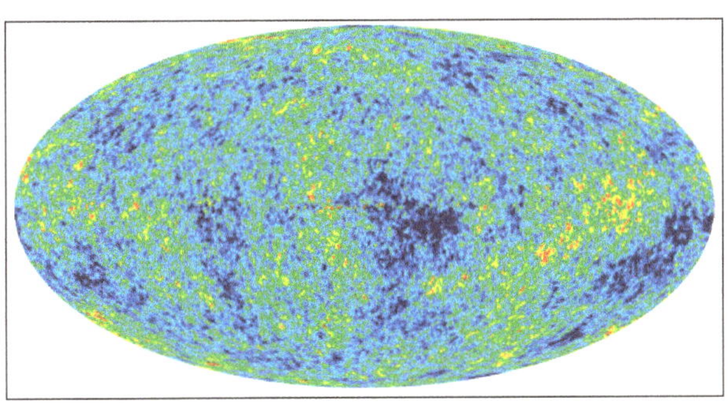

The WMAP 7-year total-intensity image of the CMB after subtraction of the dipole anisotropy and the radio-source foreground. The intensity range is only $\pm\, 200\,\mu K$ centred on the mean brightness $T_0 = 2.725\ K$.

Cosmic Neutrino Background

Cosmic neutrino background is low-energy neutrinos that pervade the universe. When the universe was one second old, it had cooled enough that neutrinos no longer interacted with ordinary matter.

Neutrinos decouple with electrons, positrons and photons at temperature $T\sim$ a few MeV and remain as such until today. Soon after the neutrino decoupling, at about T = 0.5 MeV, $e^{\pm}$ pair annihilate into photons and transfer their entropy to these photons. This transfer of entropy to photons effectively slows down the rate of decrease in photon temperature compared with that in neutrino temperature as the Universe expands. Since the entropy of neutrinos is conserved, $T_\nu = (4/11)^{1/3} T_\gamma$ and, since $T^{now}{}_\gamma = 2.7\,K$, $T^{now}{}_\nu = 1.9\,K = 1.7\times10^{-4}\,eV$. Therefore, the number density of neutrinos is given by $n_\nu = (3/22)n_\gamma = 56\ cm^{-3}$ per flavor. In the standard Big Bang model it is assumed that $n_\nu = n_{\bar{\nu}}$ and neutrinos' chemical potentials are zero. It is not assumed, in which case it is convenient to introduce an asymmetry parameter for neutrino with flavour $\alpha, \nu_\alpha : \eta_\alpha = (n_{\nu_\alpha} - n_{\bar{\nu}_\alpha})/n_\gamma = \left[\pi^2/12\zeta(3)\right]\left(\xi_\alpha + \xi^3{}_\alpha/\pi^2\right)\left(T_\nu/T_\gamma\right)^3$ where $\xi_\alpha = \mu_\alpha/T$ with μ_α being the chemical potential of ν_α. A recent study of the primordial He abundance data and WMAP data finds that the sum of the asymmetries of all neutrino flavours, taking into account the effect of neutrino oscillation with the neutrino mixing angle $sin^2\theta_{13} = 0.04\ (0.00)$, is $-0.071\ (-0.064) \le \eta_\nu \le 0.054\ (0.072)\ (90\%\ \text{C.L.})$. At the time of neutrino decoupling ($T_{\nu,dec} \sim a$ few MeV), since we know that neutrino masses are much less than a few MeV, all neutrinos are relativistic. However, as the current neutrino temperature is 1.9 K($1.7\times10^{-4}\,eV$), some of neutrinos may be non-relativistic, in which case whether neutrinos are Dirac or Majorana type is important.

Clustering of CvB

As we know that neutrino has mass, it is possible that $C\nu B$ can be trapped in gravitational potential well of some structures in the Universe such as large galaxies and clusters of galaxies when $C\nu B$ has velocity smaller than the escape velocity. For a large galaxy like Milky Way or a large cluster of galaxies the escape velocities are ~ 600 or ~ 2000 km/s, respectively. From the Maxwell–Boltzmann distribution, the average velocity of non-relativistic neutrino of mass m_ν at temperature T_ν, $<|\beta_\nu|> = \sqrt{8kT_\nu/(\pi m_\nu)} = \sqrt{4.3\times10^{-4}\text{eV}/m_\nu}$.

Therefore $<|v_\nu|> = 19{,}600\ (6200)$ km/s for $m_\nu = 0.1$ (1.0) eV. It seems that only a small fraction of neutrinos can be gravitationally trapped in a large cluster of galaxies. However, detailed simulation that takes into account either the current static mass distribution of the Milky Way (MWnow) or an estimated halo mass distribution before the formation of Milky Way through baryonic compression (NFWhalo) reveals that there may be local neutrino overdensity effect at the position of our solar system. The simulation consists of several types of weakly interacting, self-gravitating particles (cold dark matter and neutrinos) modelled as a multi-component collision less gas with the Vlasov equation and uses the density profile of cold dark matter (CDM) proposed by Navarro, Frank and White (NFWhalo). The neutrino overdensity $n_\nu/<n_\nu>$, where $<n_\nu>$ is the average neutrino density predicted by the standard Big Bang model, is with the NFWhalo model, 12 and 1.4 for m_ν = 0.6 and 0.15 eV, respectively, and with the MWnow model, 20 and 1.6 for m_ν = 0.6 eV and 0.15 eV, respectively. In the both cases (NFWhalo and MWnow) when m_ν < 0.1 eV, there is no overdensity. Overdensity values 20 and 1.4 correspond to values for the neutrino asymmetry parameter η_ν, 2.7 and 0.19, respectively.

CvB Detection Methods

Several methods to detect $C\nu B$ have been proposed and these proposed methods can be divided into three categories: (1) direct detection of coherent $C\nu B$ elastic scattering with target nuclei through momentum transfer, (2) direct detection by neutrino capture by β-decaying nuclei, and (3) indirect method by finding spectral distortion through $C\nu B$ interaction with ultra-high energy neutrinos or protons/nuclei from unknown sources.

Coherent Neutrino Elastic Scattering

There are two possible methods: use of order G_F effect and of $G^2{}_F$ effect where G_F, is the Fermi constant.

Order G_F Effect

Coherent elastic scattering occurs when the de Broglie wavelength of CνB $h / p_\nu \sim h / 4T_\nu \sim 2.4$ mm for relativistic and unclustered neutrinos, or $1.2\ mm \cdot 1\ eV / m_\nu (eV)$ for clustered non-relativistic neutrinos is much larger than inter-atomic spacing of the target material. When a neutrino with momentum p goes through coherent elastic scattering on the target and emerging with momentum p', a concept of neutrino optics can be introduced with an index of refraction $n = p' / p$ and $n - 1 \sim G_F$. Uses of neutrino optics were proposed to detect $C\nu B$ either through refraction or through total reflection as an order G_F effect. However, it was pointed out that the force induced by a linear momentum or energy exchange by neutrinos is cancelled out to order G_F when the target is in a uniform neutrino density.

The only G_F effect viable for detection of $C\nu B$ is the method proposed by Stodolsky that uses an energy split of the two spin states of non-relativistic electrons in the target immersed in $C\nu B$ for which polarized electrons in the target are needed. However, this energy split is proportional to $n_\nu - n_\nu^-$. Following Duda et al. here only their result in an optimistic case of a very large asymmetry favouring ν with $n_\nu - n_\nu^- \approx n_\nu$ is presented. Note that for overdensity of 20 $(m_\nu = 0.6\ eV)$ and of 1.4 $(m_\nu = 0.15\ eV)$ as estimated by Ringwald et al., $n_\nu - n_\nu^- =$ $0.95 n_\nu$ and $0.28 n_\nu$, respectively. The energy splits for relativistic (R)/non-relativistic (NR), clustering (C)/non-clustering (NC) and Dirac (D)/ Majorana (M) neutrinos:

$(\Delta E)_R^D = (\Delta E)_R^M = 2\sqrt{2} g_A G_F |\beta_{Earth}| n_\nu$, where $|\beta_{Earth}| = <|\beta_\nu|>$ is the velocity of the Earth relative to $C\nu B$ normalized to the speed of light in vacuum c, and g_A is 1/2 for ν_e and −1/2 for $\nu_{\mu,\tau}$. According to Ringwald et al. $<\beta_\nu> = 1.4 \times 10^{-3}$ and 2.2×10^{-3} for overdensity of 20 and 1.4, respectively. For non-clustered (NC) non-relativistic neutrinos (NR), $(\Delta E)_{NC,NR}^M = 2(\Delta E)_{NC,NR}^D \simeq 1.7\sqrt{m_\nu / (1.7 \times 10^{-4} \zeta\, \text{eV})} (\Delta E)_R^D$, and for clustered (C) neutrino non-relativistic (NR) neutrinos $(\Delta E)_{C,NR}^D = \sqrt{2} g_A G_F |\beta_{Earth}| n_\nu$ and $(\Delta E)^M{}_{C,NR} \simeq 0$. In the case of Dirac relativistic neutrinos the difference in energy ΔE between the two helicity states of an electron in the direction of $C\nu B$ induces a torque of magnitude ΔE/π. For a target of mass number A, mass M, and linear dimension R with its moment of inertia $I = MR^2 / \gamma$ (γ is a geometric factor), the acceleration due to $C\nu B$ is $a = 10^{-27} f(\gamma / 10)(100 / A)(1\ cm / R)(\beta_{Earth} / 10^{-3})\ cm / s^2$ where f is neutrino overdensity $n_\nu / <n_\nu>$. The current measurable acceleration using Cavendish type torsion balance mechanism is $\sim 10^{-13}\ cm / s^2$. With possible improvements with the current technology sensitivity down to $10^{-23}\ cm/s^2$ may be achieved. With an optimistic neutrino overdensity of 10, $\gamma = 10$, A = 100, and R = 1 cm, the expected acceleration is $10^{-26}\ cm/s^2$ to be compared with the most optimistic sensitivity to the smallest detectable acceleration of $10^{-23}\ cm/s^2$. Even in this optimistic situation, the target mass and size would need to be increased by a factor of 1000.

Order G^2_F Effect

When the Earth moves through the sea of $C\nu B$ neutrinos, a target on Earth experiences, by elastic scattering, momentum transfer from neutrinos. In the Earth's rest frame (the laboratory frame) the momentum transfer per scattering is: $<\Delta p>_R \approx \beta_{Earth}(E_\nu / c)$ for relativistic neutrinos, $<\Delta p>_{NC,NR} = \beta_{Earth}(4T_\nu / c) = <\Delta p>_R$ for non-clustering (NC) non-relativistic (NR) neutrinos, and $<\Delta p>_{C,NR} \approx \beta_{Earth} c m_\nu$ for clustered (C) non-relativistic (NR) neutrinos. The acceleration by $C\nu B$ is given by $a = \Phi_\nu (N_{A\nu} / A)\sigma_{\nu - A} < \Delta p >$ where Φ_ν is the $C\nu B$ flux, A is the mass number of the target, $N_{A\nu}$ is the Avogadro number, and $\sigma_{\nu - A}$ is the neutrino-nucleus cross section. $\sigma_{\nu - A} = G^2{}_F m^2{}_\nu / \pi \approx 10^{-56}(m_\nu / eV)^2\ cm^2$ for NR neutrinos and $= G^2{}_F E^2{}_\nu / \pi \approx 5 \times 10^{-63}(T_\nu / 1.9\ K)\ cm^2$ for R neutrinos. When the target satisfies the aforementioned condition for coherent elastic scattering, the cross section gets a coherent enhancement factor A^2. then the acceleration is: $a = \left(N_{A\nu} / A\right) n_\nu \left(G^2{}_F / \pi\right) A^2 F$ where $F = (4T_\nu)^3 \beta_{Earth}$ for R neutrinos, $F = m^2{}_\nu (4T_\nu)\beta_{Earth}$ for NC-NR neutrinos, and $F = m^3{}_\nu v^2{}_\nu$ for C-NR neutrinos. This result includes the nuclear coherent enhancement factor A^2 but does not include another possible coherent enhancement factor N_c due to the larger de Broglie wavelength of $C\nu B$ than the target size where $N_c = \left(N_{A\nu} / A\right)\rho\ (\lambda_\nu)^3$ with ρ being the density of the target. With this enhancement, Duda et al. found that the acceleration due to $C\nu B$ $a = 2 \times 10^{-34} f\rho(g\ cm^3)\ cm/s^2$ for R Dirac/Majorana neutrinos, $3 \times 10^{-28} f\left(m_\nu (eV)\right)^2 (T_\nu / 1.9\ K)^{-2}$ $\rho\left(g\ cm^3\right) cm/s^2$ for NC-NR Dirac neutrinos, and $10^{-27} f\rho(g\ cm^3)\ cm/s^2$ for C-NR Dirac neutrinos. Note that for NR Majorana neutrinos, the corresponding cross sections are reduced by $\beta^2{}_\nu \approx 10^{-6}$. Even with an optimistic scenario of $f\rho \sim 100$, $a \approx 10^{-25}\ cm/s^2$ is too small compared with the aforementioned smallest detectable acceleration of $10^{-23}\ cm/s^2$.

Cosmic Rays—CvB Scattering

It is argued that the Universe is opaque to electrons, nucleons and photons at energies higher than 10^{23} eV. A more careful consideration finds that the interaction between cosmic ray proton and CMB photon $p + \gamma_{CMB} \to \pi + N$ imposes an energy threshold known as Greisen-Zatsepin-Kuzmin (GZK) cutoff $E_{GZK} \approx 5 \times 10^{19}$ eV, beyond which cosmic ray protons do not survive. However, Weiler argues that the only particle that can survive the GKZ cut off is neutrinos and interactions of ultra-high energy cosmic ray neutrinos with C B through the Z resonance $\nu + \bar{\nu} \to Z \to p +$ any introduces a dip at certain energy in ultra-high energy neutrino flux. If these neutrinos come from sources with the redshift z = 3.5, this dip occurs at 9×10^{19} eV. Depending on the three neutrino masses and the value of redshift parameter z of the sources of these ultra-high energy neutrinos, there may be possibly three dips in energy spectrum of cosmic ray neutrinos if these sources exist. The existence of these neutrino sources can produce cosmic rays beyond the GZK cut off through the Z resonance (Z-burst). However, although the AGASA experiment claimed that they detected cosmic ray events above the GZK cutoff; neither the HiRes Experiment nor the Auger experiment confirmed such events.

Another interaction of cosmic rays with $C\nu B$ to detect $C\nu B$ was proposed by Wigman. He proposes to explain, in terms of inverse β-decay interactions such as $p+\bar{\nu}_e \rightarrow n+e^+$, the changes in the power index of cosmic rays spectrum around $10^{15.3}$ eV (the first knee) observed by the CASA-BLANCA experiment from $n = 2.72 \pm 0.02$ to $n = 2.95 \pm 0.02$, and also at the second knee around $10^{17.5}$ eV from $n = 3.01 \pm 0.06$ to $n = 3.27 \pm 0.02$ observed by the Fly's Eye experiment. In the interaction $p+\bar{\nu}_e \rightarrow n+e^+$ where $\bar{\nu}_e$ is $C\nu B$, the centre of mass energy $E_{CM} \approx \sqrt{m_p^2 + 2E_p m_\nu}$ should be greater than the sum of proton and neutron masses $m_p + m_n$. This leads to the threshold proton energy for the interaction at $1.695 \times 10^{15} / \left(m_\nu / (1\ eV)\right)\ eV$. If the energy of the knee is, combining several experimental data, at $(3 \pm 1) \times 10^{15}$ eV and the cause of the first knee is the inverse β-decay interaction by proton, it is consistent with the neutrino mass $m_{\nu_e} = 0.5 \pm 0.2\ eV/c^2$.

Although evidence of high energy cosmic ray neutrinos has finally emerged as the Ice Cube experiment shows, there is only one experimental result that suggests that there are cosmic rays beyond the GZK cut off, which has not been confirmed by the HiRes and Auger experiment. Various mechanisms in addition to the inverse β-decay theory have been proposed to explain the existence of the knees in cosmic ray spectrum. However, it is not clear which explanation is the correct model for the knees.

Cosmic Infrared Background

The Cosmic Infrared Background (CIRB) is the radiation from stars in many faint galaxies. It is what is left over after emission from our Solar System and our Galaxy has been subtracted away. Here are pictures of the sky before and after this foreground subtraction. The near-infrared (wavelengths near 2-3 microns) and optical (wavelengths near 500 nm) part of this extragalactic background light is just starlight redshifted into the infrared. But some starlight is absorbed by dust and re-emitted in the far-infrared (wavelengths near 100 microns). The figure below shows the cosmic near-infrared background in red in the center right, Rebecca Bernstein's optical cosmic background in blue on the far right, the FDS 60 and 100 micron values in blue, the cosmic microwave background in grey on the far left.

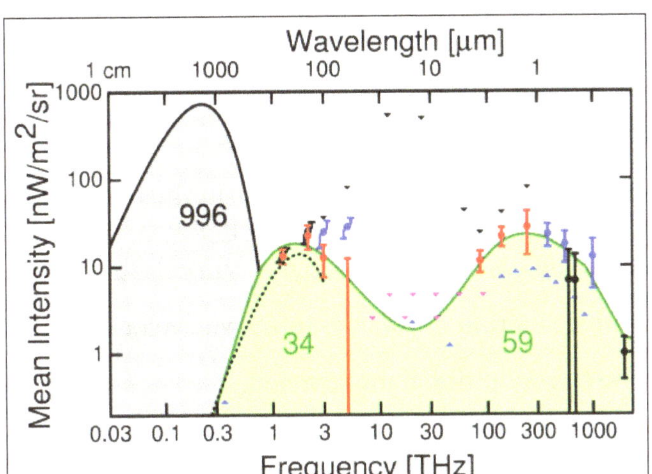

The CMB is by far the largest of these radiation fields, with a total intensity of 996 nW/m²/sr. The cosmic far-IR background, which was announced in January 1998, has a total intensity of 34 nW/

m^2/sr, while the cosmic near-IR background has a total intensity when combined with the optical background of slightly less than 60 nW/m^2/sr. Together these IR backgrounds add up to about 9% of the CMB's total intensity.

The black data points between 1 and 300 microns on this graph come from the DIRBE experiment on the COBE satellite. The blue lower limit symbols are based on integrating galaxy counts, while the purple upper limit symbols are based on limits on photon-photon collisions from gamma-ray astronomy. The curve is the Lambda CDM model with the Salpeter IMF, multiplied by a factor of 1.84, and with modifications for wavelengths longer than 300 microns to fit the FIRAS distortion limits, and for wavelengths shorter than 0.8 microns to fit the optical and UV data.

Hubble's Law

Hubble's law statement is given as:

The velocity of the galaxy which is also known as the redshift is directly proportional to its distance. Hubble's law is also known as Hubble-Lemaitre law, studied in physical cosmology. Only with the help of Hubble's law the universe can expand and fulfil the cosmological principle.

Hubble's Law Formula

Hubble's law formula is:

$$v = H_0 d$$

Where,

- v is the velocity of the galaxy in km/s.
- H_0 is the Hubble constant in km/s/Mpc.
- d is the distance of the galaxy in Mpc.

Limitations of Hubble's Law

By determining the shift in observed light into redshift in the spectrum, one can determine the distance of the galaxy from us using Hubble's equation after measuring the recession velocity. Following are the limitations of Hubble's law which makes the measurement challenges:

- Because of the intrinsic motion of galaxies, observed velocity gets influenced.
- Galaxy orbiting due to gravitational movements.

What is Hubble Constant?

Hubble constant is defined as the unit of measurement, which is used for describing the expansion of the universe. Hubble's constant H is 160 km/s per million-light-years.

What is Redshift?

Redshift is the phenomenon in which an object's wavelength increases due to electromagnetic radiation. Blueshift is opposite to redshift where the energy increases due to shorter wavelengths which are also known as negative redshift. The main reasons for redshift are as follows:

- Doppler Effect i.e.; the movement of objects either closer or apart from each other in space.
- Strong gravitational force leads to gravitational redshift.
- Cosmological redshift is an expansion of space such that objects are separated without the change in their position.

Interpretation of Redshift and Distance

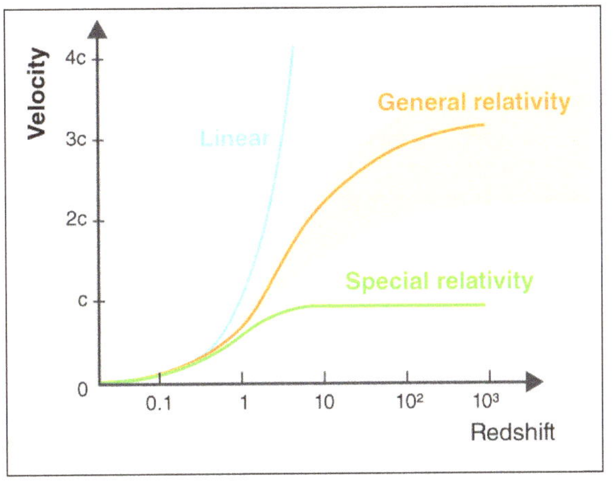

Above is the linear relationship between the redshift and distance and the mathematical expression for Hubble's law as follows:

$$v = H_o D$$

Where,

- v is the recessional velocity in km/s.
- H_o is the Hubble's constant.
- D is the proper distance between the galaxy and the observer.

Redshift Formula

The redshift formula is given as:

$$z = \frac{\Delta\lambda}{\lambda}$$

Where,

- z is the redshift.

- $\Delta\lambda$ is the shift in wavelength in the spectra.
- λ is the wavelength.

Units Derived from Hubble Constant

There are three units that are derived from Hubble constant, and they are:

- Hubble Time: The Hubble time is defined as the age taken by the universe to expand if the expansion was linear and in different real age. This is dependent on the dimensionless factor which is dependent on the mass and energy of the content in the universe.
    - Hubble time is given as follows:

      tH = 1/H_0 = 1/67.8 (km/s)/Mpc = 4.55.1017s = 14.4 billion years.
- Hubble Length: The Hubble length is defined as the product of the speed of light and the Hubble time. It is also known as the Hubble distance. The product obtained is equivalent to 14.4 billion light-years. cH_0^{-1} is the Hubble length.
- Hubble Volume: The Hubble volume is defined as the volume of the observable universe. It is also defined as the volume of sphere whose radius is cH_0^{-1} or a square with side cH_0^{-1}.

Friedmann Equation

Newtonian Derivation of the Friedmann Equation

Consider an isolated sphere of radius R_s and mass M_s, in uniform, isotropic expansion (Hubble flow).

The equation of motion for $R_s(t)$ can be obtained from the gravitational acceleration at the outer edge of the sphere:

$$\frac{d^2R_s}{dt^2} = -\frac{GM_s}{R_s^2|(t)}.$$

Multiplying both sides by dR_s / dt and integrating converts this "acceleration equation" to an "energy equation":

$$\frac{1}{2}\left(\frac{dR_s}{dt}\right)^2 = \frac{GM_s}{R_s(t)} + U.$$

Mathematically, U is just a constant of integration, but physically it corresponds to the total energy per unit mass at the surface of the expanding sphere, i.e., the sum of the kinetic energy per unit mass and the gravitational potential energy per unit mass. If U > 0, then the expanding sphere has positive total energy and will expand forever (the r.h.s. will always be positive).

If U < 0, then the sphere has negative total energy and will eventually recollapse (the r.h.s. will eventually become zero).

Write the radius in the form,

$$R_s(t) = a(t) r_s,$$

Where r_s is the "co-moving" radius of the sphere, equal to the physical radius at the epoch when a (t) = 1.

with,

$$M_s = \frac{4\pi}{3} \rho(t) R_s^3(t),$$

the energy equation becomes,

$$\frac{1}{2} r_s^2 \dot{a}^2 = \frac{4\pi}{3} G r_s^2 \rho(t) a^2(t) + U.$$

Dividing each side by $r_s^2 a^2 / 2$ yields:

$$\left(\frac{\dot{a}}{a}\right)^2 = \frac{8\pi G}{3} \rho(t) + \frac{2U}{r_s^2 a^2(t)}.$$

Since $\rho(t) \propto 1/a^3(t)$, we see that if U is negative the r.h.s. of this equation will eventually hit zero, after which the expansion reverses.

Although this derivation describes an isolated sphere, Newton's "iron-sphere" theorem tells us that it should also describe any spherical volume of a homogeneous and isotropic universe, since the gravitational effects of a spherically symmetric external matter distribution cancel. A generalization of the "iron-sphere" theorem turns out to hold in GR as well.

The Friedmann Equation in GR

A proper derivation of the Friedmann equation begins by inserting the Friedmann-Robertson-Walker metric into the Einstein Field Equation. Since GR yields the Newtonian limit, we should expect the small scale behavior to resemble that of our Newtonian derivation above, and it does, with two important changes.

First, the mass density $\rho(t)$ is replaced by the total energy density $\in(t)/c^2$, which includes rest mass energy and other forms of energy (e.g., energy of photons, or thermal energy of atoms). (Often this total energy density is just written as $\rho(t)$ and understood to include all contributions, not just rest mass. I will try to stick with Ryden's more pedagogical notation here.) Second, the "potential energy" term is intimately tied to the curvature of space.

The GR form of the Friedmann Equation is:

$$\left(\frac{\dot{a}}{a}\right)^2 = \frac{8\pi G}{3} \frac{\varepsilon(t)}{c^2} - \frac{kc^2}{R_0^2} \frac{1}{a^2(t)},$$

Where R_0 is the present value of the curvature radius and k = +1, 0, or -1 is the curvature index in the FRW metric.

While the precise form of the last term is not obvious without a GR derivation, it makes reasonable sense that positive space curvature is associated with stronger gravity and thus with negative "binding energy."

If k ≤ 0 and the energy density is positive, then the r.h.s. is always positive, and an expanding universe continues to expand forever.

If matter is the dominant form of energy, then dilution implies $\in(t) \propto 1/a^3(t)$. f k = +1, then the r.h.s. must eventually reach zero, after which the expansion will reverse.

Thus, positive space curvature corresponds to a bound universe. However, a form of energy for which $\in(t)$ falls more slowly than $1/a^2(t)$, such as a cosmological constant, can change this automatic correspondence.

The Critical Density and the Density Parameter, Ω

Substituting $H(t) = \dot{a}/a$ allows us to write the Friedmann equation in terms of the Hubble parameter,

$$H^2(t) = \frac{8\pi G}{3}\frac{\varepsilon(t)}{c^2} - \frac{kc^2}{R_0^2}\frac{1}{a^2(t)}$$

From this equation, we can see that space is flat (k = 0) if the mean density of the universe equals the critical density,

$$\rho_c(t) = \frac{\varepsilon_c(t)}{c^2} = \frac{3H^2(t)}{8\pi G}.$$

The two cosmological equations most worth memorizing are $H = \dot{a}/a$ and this definition of the critical density. Together they are the Friedmann equation for a flat universe. The present day value of the critical density is,

$$\rho_{c,0} = \frac{\varepsilon_{c,0}}{c^2} = \frac{3H_0^2}{8\pi G}$$

$$= 9.2\times10^{-30}\,g\,\mathrm{cm}^{-3}\left(\frac{H_0}{70\ \mathrm{km\ s^{-1}Mpc^{-1}}}\right)^2$$

$$= 1.4\times10^{11}\,M_\odot Mpc^{-3}\left(\frac{H_0}{70\ \mathrm{Km\ s^{-1}Mp\ c^{-1}}}\right).$$

Cosmologists frequently describe the energy density of the universe in terms of the density parameter,

$$\Omega \equiv \frac{\varepsilon}{\varepsilon_c} = \frac{\varepsilon}{c^2}\times\frac{8\pi G}{3H^2},$$

the ratio of the total energy density ε to the critical energy density.

Substituting this definition into the Friedmann equation yields,

$$H^2 = \Omega H^2 - \frac{kc^2}{R_0^2 a^2(t)} \Rightarrow 1 - \Omega(t) = -\frac{kc^2}{H^2(t) a^2(t) R_0^2}.$$

If $\Omega = 1$, then it equals one at all times, since the r.h.s. of this equation always vanishes.

In other cases, the value of Ω changes with time, but if $\Omega > 1$ it is always > 1, and if $\Omega < 1$ it is always < 1, because the r.h.s. cannot change sign. At the present day, we can solve this equation to get,

$$R_0 = \frac{c}{H_0} |1 - \Omega_0|^{-1/2}.$$

If Ω_0 is very close to one, then the curvature radius is large compared to the Hubble radius c/H_0, and curvature effects on this scale are small.

Evolution of Energy Density: The Fluid Equation

The Friedmann equation determines a (t) if we know H_0 and the energy density $\varepsilon(t)$ as a function of time.

(By comparing ε_0 to the critical density, we can determine whether k = + 1, 0, or −1, and we can use our last equation to determine R_0.)

If the only energy contribution is from non-relativistic matter, then $\varepsilon(t) = \varepsilon_0 (a / a_0)^{-3}$, since expansion of the universe simply dilutes the density of particles. For the more general case, let's turn to the first law of thermodynamics,

$$dE = -PdV + dQ,$$

the change of internal energy of a volume of fluid is the sum of PdV work and added heat.

The expansion of a homogeneous universe is adiabatic, as there is no place for "heat" to come from, and no "friction" to convert energy of bulk motion into random motions of particles.

(There is a caveat to this statement: when particles annihilate, such as electrons and positrons, this adds heat and makes the expansion temporarily non-adiabatic. This matters at some specific epochs in the very early universe).

Therefore,

$$dE + PdV = 0 \Rightarrow \dot{E} + P\dot{V} = 0.$$

For a sphere of co-moving radius $r_s, V = \frac{4\pi}{3} r_s^3 a^3(t)$,

$$\dot{V} = \frac{4\pi}{3} r_s^3 \left(3s^2\dot{a}\right) = V\left(3\frac{\dot{a}}{a}\right),$$

And $E = V\varepsilon$,

Therefore,

$$\dot{E} = V\dot{\varepsilon} + \dot{V}\varepsilon = V\left(\dot{\varepsilon} + 3\frac{\dot{a}}{a}\varepsilon\right).$$

Together with $\dot{E} + P\dot{V} = 0$, we get,

$$V = \left(\dot{\varepsilon} + 3\frac{\dot{a}}{a}\varepsilon + 3\frac{\dot{a}}{a}P\right) = 0$$

and thus,

$$\dot{\varepsilon} + 3\frac{\dot{a}}{a}\left(\varepsilon + P\right) = 0.$$

This fluid equation describes the evolution of energy density in an expanding universe. To solve this equation, we need an additional equation of state relating P and ε. Suppose we write this in the form,

$$P = w\varepsilon.$$

In principle, w could change with time, but we will assume that any time derivatives of w are negligible compared to time derivatives of ε. This is reasonable if the equation of state is determined by "microphysics" that is not directly tied to the expansion of the universe. The fluid equation then implies:

$$\frac{\dot{\varepsilon}}{\varepsilon} = -3\left(1+w\right)\frac{\dot{\varepsilon}}{a},$$

With solution,

$$\frac{\varepsilon}{\varepsilon_0} = \left(\frac{a}{a_0}\right)^{-3(1+w)}.$$

For non-relativistic matter,

$$w = P/\varepsilon \sim \frac{mv_{\text{th}}^2}{mc^2} \sim \frac{v_{\text{th}}^2}{c^2} \ll 1,$$

Where, v_{th} is the thermal velocity of particles.

To a near-perfect approximation, w = 0, implying $\varepsilon \propto a^{-3}$, in line with our simple dilution argument.

For radiation (i.e., photons), w = 1/3, implying $\varepsilon \propto a^{-4}$.

This behaviour also follows from a simple argument: the number density of photons falls as $n \propto a^{-3}$, and the energy per photon falls as $h\nu \propto a^{-1}$ because of cosmological redshift. The fluid equation will lead us to some less obvious conclusions when we consider "dark energy".

The Acceleration Equation

If we multiply our standard version of the Friedmann equation by a^2, we get,

$$\dot{a}^2 = \frac{8\pi G}{3c^2}\varepsilon a^2 - \frac{kc^2}{R_0^2}.$$

Take the time derivative,

$$2\dot{a}\ddot{a} = \frac{8\pi G}{3c^2}\left(\dot{\varepsilon}a^2 + 2\varepsilon a\dot{a}\right),$$

Divide by $2\dot{a}a$,

$$\frac{\ddot{a}}{a} = \frac{4\pi G}{3c^2}\left(\dot{\varepsilon}\frac{a}{\dot{a}} + 2\varepsilon\right),$$

And substitute from the fluid equation,

$$\dot{\varepsilon}\frac{a}{\dot{a}} = -3(\varepsilon + P)$$

To get,

$$\frac{\ddot{a}}{a} = -\frac{G}{c}(\quad +3\quad).$$

We see that if ε and P are positive, the expansion of the universe decelerates. Higher P produces stronger deceleration for given ε, e.g., a radiation-dominated universe decelerates faster than a matter-dominated universe.

The appearance of $\varepsilon / c^2 + 3P / c^2$ is a specific example of a more general phenomenon in GR: pressure appears in the stress-energy tensor, and it therefore has a gravitational effect. (It must, because the division between energy density and pressure depends on the state of motion of the observer.) With the stress-energy tensor of an ideal fluid, the Newtonian limit of GR yields a "Poisson equation":

$$\nabla^2\Phi = 4\pi G\left(\rho + 3P / c^2\right).$$

The Cosmological Constant, Vacuum Energy and Cosmic Acceleration

A negative energy density would be pretty bizarre. Negative pressure sounds bizarre too, but it's not quite as crazy. For a fluid to have negative pressure means that it has tension — it takes work to expand the fluid, instead of taking work to compress it.

Suppose the universe is pervaded by a form of energy that is constant density, in space and time. Such energy could conceivably arise as a consequence of the quantum vacuum, which is not really empty but filled with "virtual" particles.

Expanding a volume increases its energy, so that energy must have come from doing P dV work on the volume, which implies that the pressure must be negative (the volume "resists stretching.") In terms of our fluid equation, we see that ε = const. implies w = −1, and thus P = $-\varepsilon$.

From the acceleration equation, we see that any fluid with P < $-\varepsilon$ /3 causes acceleration of the universe, instead of deceleration. Such a fluid has, in GR, repulsive gravity. This discussion makes cosmological acceleration sound almost reasonable. The problem is that no one knows how to calculate the quantum vacuum energy from first principles.

In the absence of a complete theory of quantum gravity, the most reasonable guess is that,

$$\varepsilon_{\text{vac}} \sim E_p / l_p^3,$$

Where $l_p = \left(\hbar G / c^3\right)^{1/2} = 1.6 \times 10^{-35}\,\text{m}$ is the Planck length and $E_p = \left(\hbar c^5 / G\right)^{1/2} = 2.0 \times 10^9\,\text{J}$ is the Planck energy. This estimate exceeds the critical density by 120 orders of magnitude.

Since the only "natural" number close to 10^{-120} is zero, the general expectation at least until the mid-1990s was that a "correct" calculation of quantum vacuum energy would produce cancellations that make the answer exactly zero. It's possible that vacuum energy really does have the value required to produce the observed cosmic acceleration.

It's also possible that the fundamental vacuum energy is zero, and that acceleration of the Universe is caused by some other negative pressure fluid, or by a breakdown of GR on cosmological scales.

Einstein introduced the "cosmological constant" Λ with a different conception, as a modification of the curvature term of the Field Equation rather than an additional contribution to the stress-energy tensor.

However, the gravitational effects of Einstein's cosmological constant are identical to those of a form of energy with P = $-\varepsilon$, which remains constant in space and time as the universe expands. Einstein needed the "repulsive gravity" of the Λ-term to allow a static universe, counteracting the gravitational attraction of matter. He abandoned the idea when the universe was found to be expanding. But today we need a cosmological constant, or something like it, to explain how the universe can accelerate.

Future of an Expanding Universe

The Big Bang model describes the current universe as expanding. Based on the latest observations, we conclude that the expansion will continue forever. What will happen to Earth, the Solar System, and the rest of the universe as time goes by?

The Solar System

Our Solar System has been around for about 4.5 billion years. The Sun is a star of relatively low mass, which means that it burns its nuclear fuel relatively slowly. In the core of the Sun, hydrogen atoms are fusing to form helium. The energy released by this reaction pushes the solar material outwards, supporting it against gravity. Eventually, the radiation makes its way to the surface, then flies into space as (mostly) visible light.

In about 6 billion years, the Sun will have converted almost all the hydrogen in its core into helium. Eventually, the helium in the core will start to fuse together to form carbon. This reaction takes place at much higher pressures and temperatures than hydrogen fusion, and releases much more energy. The extra energy generated in the core will push the outer layers of the Sun far, far beyond their current position. As they are pushed outwards, the outermost layers of the Sun's atmosphere will cool down from the current 5700 degrees Kelvin to perhaps 4000 or 3500 degrees Kelvin. The Sun will swell up and become cooler and redder: a red giant.

The radius of the Sun will grow very large during this phase of its evolution. Its outer layers will certainly engulf the planet Mercury, possibly Venus, and maybe even the Earth. But, even if the Earth escapes direct contact with the solar atmosphere, it will be doomed nonetheless: the solar luminosity will increase by several thousand as it becomes a red giant, heating the Earth to a temperature of about 2000 Kelvin (about 3000 Fahrenheit). The atmosphere will fly off into space, the oceans will boil away, and much of the continents will melt.

The red giant phase of life is a brief one: after only about 1 billion years, the Sun will deplete all the helium in its core. It does not have sufficient mass to drive pressures and temperatures high enough to fuse carbon into heavier elements. Instead, with no source of energy at its centre, the

Sun's luminosity will fall drastically. Its swollen outer layers will be ejected into space, forming a planetary nebula, and leaving a small, dense remnant behind: a white dwarf.

Although initially very hot (40,000 Kelvin or so), the white dwarf will cool as it radiates away its internally energy. After several billion years, its temperature will have decreased to a few thousand degrees Kelvin and its light will have dimmed to a dull red. The Earth's surface will cool and again solidify, then continue to cool far past our current temperature. Ten billion years from now, what remains of the Earth will be frozen, more than a hundred degrees below zero. Any oxygen or nitrogen gas on the surface (emitted by volcanic activity, perhaps) will be frozen as a frost on the ground.

The Stelliferous Age

Stelliferous means "star-bearing". The current universe is indeed stelliferous: we see new stars being created in galaxies almost everywhere we look. Stars are, in fact, one of the dominant features of the universe right now.

When we look at our Milky Way galaxy (or other galaxies) from an overall point of view, we find that much of its light - perhaps more than half - comes from a relatively few stars: massive, blue, young stars, less than one billion years old, emit huge amounts of ultraviolet, visible, and infrared radiation. They are often grouped together into star-forming regions, sometimes lighting up gas in their vicinity to form HII regions.

But these massive stars run through their store of nuclear fuel very quickly. The rate at which fusion occurs in a stellar core depends very sensitively on the temperature and pressure. Low-mass stars, with lower temperatures and pressures, consume their nuclear fuel much more slowly than high-mass stars. Even though the low-mass stars start with much less fuel, they continue to burn it for a much longer time.

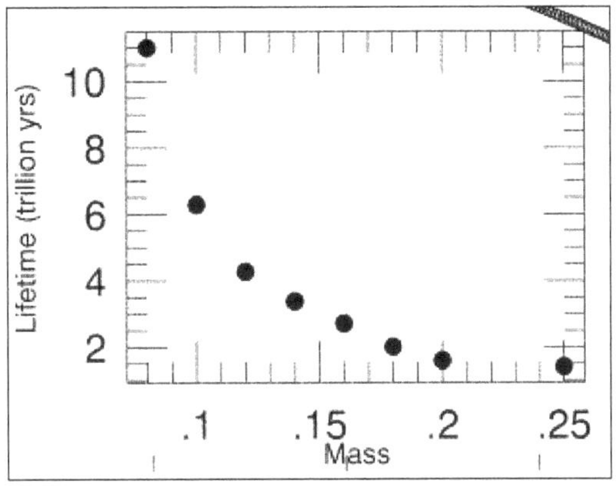

Initial mass of star	Lifetime
10 solar masses	0.02 billion years
5	0.07
2	1
1	10
0.5	
0.2	1000
0.08	10000

Objects will masses less than about 0.08 solar do not create high enough temperatures or densities in their cores to permit fusion - so they do not become stars. The stars of lowest mass are small, cool, and dim: we call them red dwarfs. Their luminosities are so low - less than one per cent of the Sun's - that only planets with very small orbits could be warm enough to have liquid water. The most frugal of them may continue to burn hydrogen into helium for another 10,000 billion, or 10 trillion, years.

But eventually, even the red dwarfs will run out of fuel. They will cool off and become black dwarfs: chunks of helium and heavier elements bound together by gravity, radiating only tiny amounts of energy as they very slowly settle to a final shrunken state. After about 10^ (14), or 100 trillion, years, the stars are dead.

The Degenerate Age

After the Stelliferous age, the universe is very different, for two reasons. First, stars are no longer shining. Second, galaxies which are not gravitationally bound to each other have been carried far away by the expansion of the universe. After several passes between our Milky Way and Andromeda, the two galaxies will merge. After trillions of years, all the galaxies in our Local Group will have coalesced into one big object. But all the other galaxies in the universe will have continued to move farther and farther away: by 10^ (14) years in the future, even our nearest unbound neighbours will be more than one million Mega parsecs away from us.

The dominant forms of matter are now mostly dead stars, in several forms:

- Brown dwarfs: Objects not massive enough to form stars (about the size of Jupiter).
- White dwarfs: The dense, cooling remains of low-mass stars (about the size of the Earth).
- Neutron stars: The very dense remains of high-mass stars which ran out of fuel and suffered core-collapse (about the size of New York City).
- Black holes: The very very dense remains of high-mass stars whose cores were more than about three solar masses at the time of core collapse (about the size of Rochester).

The atoms in all these objects are unlike those in ordinary matter, due to their high density. Physicists call this sort of material degenerate, and that term gives its name to this Age. Since these objects are not burning any fuel, they do not radiate light. The Degenerate Age is dark. Over the course of this Age, gravity causes two competing effects:

- Chance encounters between stellar remnants in our merged "Local Galaxy" cause some of the stars to fly off into intergalactic space, and others to fall deeper into the gravitational well of the "Local Galaxy".

  If a remnant happens to pass close to a black hole, it may be torn apart. The resulting material may form an accretion disk around the black hole, spiraling inwards and emitting radiation as it goes.

  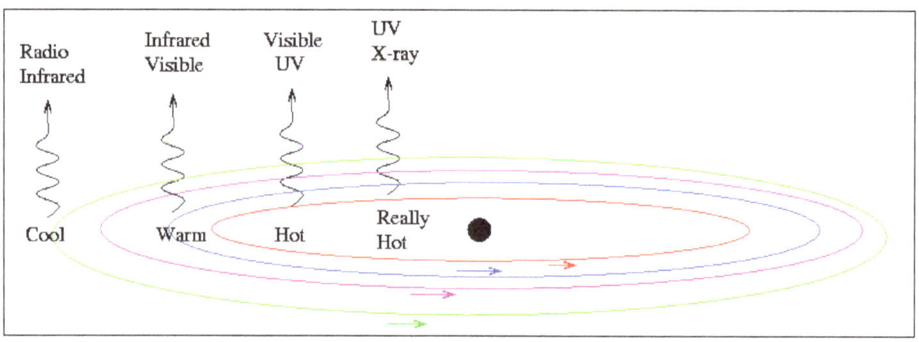

  The accretion process might last a few years, but will inevitably die away when all the material has fallen into the black hole.

- Objects in orbit around each other - binary stars, or planets around a single star - emit very small amounts of gravitational radiation. Over ordinary timescales, the loss of energy is negligible; but during the Degenerate Era, over periods of quadrillions or quintillions of years, this tiny "leak" causes orbits to shrink. Eventually, all objects in close orbits merge, destroying any planetary systems or double stars.

If two white dwarfs with a combined mass of more than about 1.4 solar masses happen to merge, the resulting object undergoes runaway thermonuclear fusion. The result is a Type Ia supernova. Very, very rarely, the darkness of the Degenerate Age is dispelled for a few weeks while a supernova explodes. At a time of 10^ (25) years, the material in the "Local Galaxy" consists of isolated stellar remnants and black holes. Everything is cold and dark.

Some theories of physics predict that protons - one of the fundamental particles which make up matter - ought to be unstable over very long timescales. If those theories are correct, then after 10^(30) years or so, the protons in brown dwarfs, white dwarfs, and neutron stars will start to decay. The decaying protons turn into radiation, which flies off into the dark. Eventually, all these remnants disintegrate, leaving behind nothing but electrons, positrons, neutrinos, and radiation. Only the black holes would be left.

The Black Hole Age

If protons really do decay, then after about 10^(30) years, black holes will be the only collections of matter which remain. Do they last forever, slowly orbiting each other in the isolated, coalesced galaxies?

No! Black holes, it turns out, do emit a very, very small amount of radiation. Classically, nothing can escape their gravitational pull, not even the light; but the rules of quantum mechanics permit a process by which photons or other elementary particles may - on very rare occasions - be emitted from the black hole. This is sometimes called "Hawking radiation." Under ordinary circumstances, it is minimal: the radiation emitted by a black hole of one solar mass corresponds to a temperature of 0.0000001 degrees Kelvin.

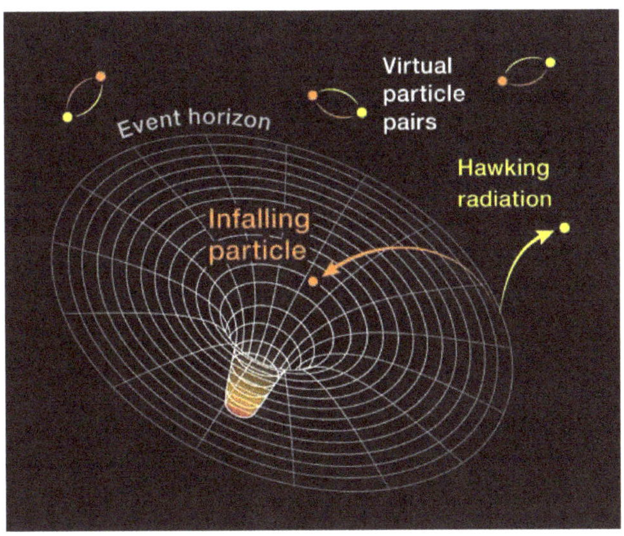

Nonetheless, during the Black Hole Age, there is plenty of time for this miniscule amount of radia tion to carry away a significant portion of a black hole's mass. Moreover, as a black hole loses mass, its temperature and luminosity rise, so the process speeds up. Eventually, even the most massive black holes must evaporate due to this radiation.

Mass of black hole	Time to evaporate
7 × 10^(22) kg (Moon)	10^(43) years
6 × 10^(24) kg (Earth)	10^(49) years
2 × 10^(30) kg (Sun)	10^(66) years
2 × 10^(36) kg (million Sun)	10^(84) years
2 × 10^(42) kg (galaxy)	10^(102) years

As its mass decreases, the amount of radiation emitted by a black hole increases. During the last few seconds of its life, an evaporating black hole emits a burst of light, X-rays, and gamma rays. The cold, dark space of the Black Hole Age will very, very, very occasionally be lit up as a black hole ends its life. After about 10^ (100) years, all the black holes are gone.

The Dark Age

After all the black holes have evaporated, (and after all the ordinary matter made of protons has disintegrated, if protons are unstable), the universe will be nearly empty. Photons, neutrinos, electrons and positrons will fly from place to place, hardly ever encountering each other. It will be cold, and dark, and there is no known process which will ever change things.

2

Understanding Cosmology

Cosmology is a branch of astronomy concerned with the studies of the physical universe including origin and evolution of the universe. The chapter elaborates big bang theory, cosmological spacetime, structure formation, cosmological components and cosmological constant. This chapter has been carefully written to provide an easy understanding of the varied facets of cosmology.

Cosmology is a branch of astronomy that involves the origin and evolution of the universe, from the Big Bang to today and on into the future. According to NASA, the definition of cosmology is "the scientific study of the large scale properties of the universe as a whole".

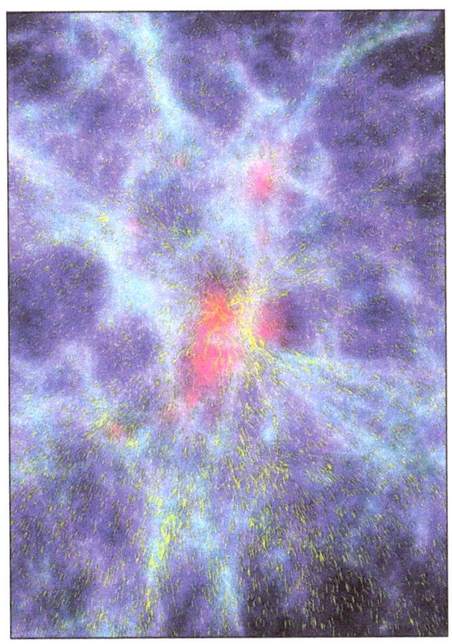

Snapshot from a computer simulation of the formation of large-scale structures in the universe, showing a patch of 100 million light-years and the resulting coherent motions of galaxies flowing toward the highest mass concentration in the center.

Cosmologists puzzle over exotic concepts like string theory, dark matter and dark energy and whether there is one universe or many (sometimes called the multiverse). While other aspects astronomy deal with individual objects and phenomena or collections of objects, cosmology spans the entire universe from birth to death, with a wealth of mysteries at every stage.

Humanity's understanding of the universe has evolved significantly over time. In the early history of astronomy, Earth was regarded as the centre of all things, with planets and stars orbiting it. In the 16th century, Polish scientist Nicolas Copernicus suggested that Earth and the other planets in the solar system in fact orbited the sun, creating a profound shift in the understanding of the cosmos. In the late 17th century, Isaac Newton calculated how the forces between planets — specifically the gravitational forces — interacted.

The dawn of the 20th century brought further insights into comprehending the vast universe. Albert Einstein proposed the unification of space and time in his General Theory of Relativity. In the early 1900s, scientists were debating whether the Milky Way contained the whole universe within its span, or whether it was simply one of many collections of stars. Edwin Hubble calculated the distance to a fuzzy nebulous object in the sky and determined that it lay outside of the Milky Way, proving our galaxy to be a small drop in the enormous universe. Using General Relativity to lay the framework, Hubble measured other galaxies and determined that they were rushing away from us, leading him to conclude that the universe was not static but expanding.

In recent decades, cosmologist Stephen Hawking determined that the universe itself is not infinite but has a definite size. However, it lacks a definite boundary. This is similar to Earth; although the planet is finite; a person traveling around it would never find the "end" but would instead constantly circle the globe. Hawking also proposed that the universe would not continue on forever but would eventually end.

Cosmological Missions and Instruments

Launched in November 1989, NASA's Cosmic Background Explorer (COBE) took precise measurements of radiation across the sky. The mission operated until 1993. Although NASA's Hubble Space Telescope is probably best known for its astounding images, a primary mission was cosmological. By more accurately measuring the distances to Cepheid variables, stars with a well-defined ratio between their brightness and their pulsations, Hubble helped to refine measurements regarding how the universe is expanding. Since its launch, astronomers have continued to use Hubble to make cosmological measurements and refine existing ones.

"If you put in a box all the ways that dark energy might differ from the cosmological constant, that box would now be three times smaller," cosmologist Adam Riess of the Space Telescope Science Institute said in a statement. "That's progress, but we still have a long way to go to pin down the nature of dark energy."

NASA's Wilkinson Microwave Anisotropy Probe (WMAP) was a spacecraft that operated from 2001 to 2010. WMAP mapped tiny fluctuations in the cosmic microwave background (CMB), the ancient light from the early universe, and determined that ordinary atoms make up only 4.6 per cent of the universe, while dark matter makes up 24 per cent.

"Lingering doubts about the existence of dark energy and the composition of the universe dissolved when the WMAP satellite took the most detailed picture ever of the cosmic microwave background," said cosmologist Charles Seife in the journal Science.

The European Space Agency's Planck space mission ran from 2009 to 2013 and continued the study of the cosmic microwave background. The ESA is currently developing the Euclid mission, which should fly by the end of the decade. Euclid will study dark matter and dark energy with greater precision, tracing its distribution and evolution through the universe.

"At the heart of the mission is one of the billion pound questions of physics."

What Came before the Big Bang?

Because of the enclosed and finite nature of the universe, we cannot see "outside" of our own universe. Space and time began with the Big Bang. While there are a number of speculations about the existence of other universes, there is no practical way to observe them, and as such there will never be any evidence for (or against!) them.

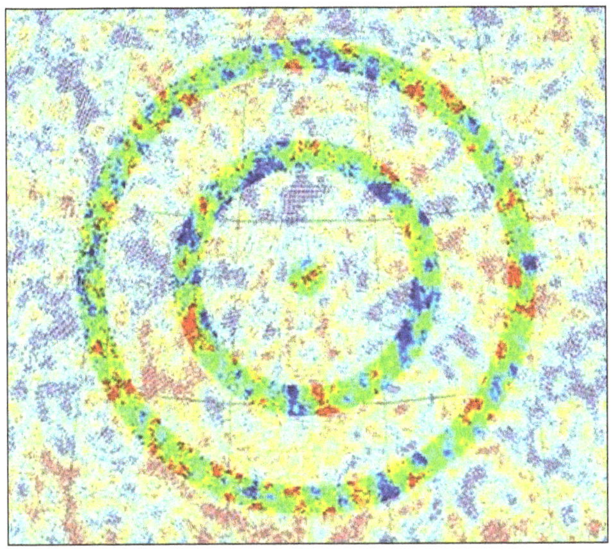

Some researchers think concentric ring patterns in measurements of the cosmic microwave background are evidence of a universe that existed before our own was born in the Big Bang.

Where did the Big Bang happen?

The Big Bang did not happen at a single point but instead was the appearance of space and time throughout the entire universe at once.

If other galaxies all seem to be rushing away from us, doesn't that place us at the center of the universe?

No, because if we were to travel to a distant galaxy, it would seem that all surrounding galaxies were similarly rushing away. Think of the universe as a giant balloon. If you mark multiple points on the balloon, then blow it up, you would note that each point is moving away from all of the others, though none are at the centre. The expansion of the universe functions in much the same way.

How Old is the Universe?

According to data released by the Planck team in 2013, the universe is 13.8 billion years old, give or take a hundred million years or so. Planck determined the age after mapping tiny temperature fluctuations in the CMB.

"Patterns over huge patches of sky tell us about what was happening on the tiniest of scales in the moments just after our universe was born," said Charles Lawrence, the U.S. project scientist for Planck, in a statement.

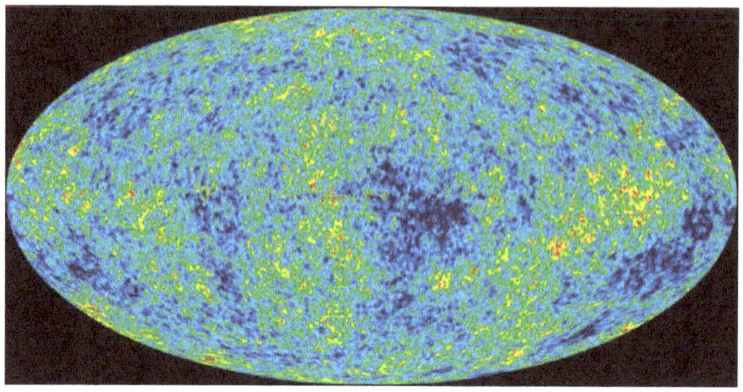

Will the Universe end? If so, then how?

Whether or not the universe will come to an end depends on its density — how spread out the matter within it might be. Scientists have calculated a "critical density" for the universe. If its true density is greater than their calculations, eventually the expansion of the universe will slow and then, ultimately, reverse until it collapses. However, if the density is less than the critical density, the universe will continue to expand forever.

The post-Big Bang universe was composed predominantly of hydrogen, with a little bit of helium thrown in for good measure. Gravity caused the hydrogen to collapse inward, forming structures. However, astronomers are uncertain whether the first massive blobs formed individual stars that later fell together via gravity, or the mass came together in galaxy-sized clumps that later formed stars.

Duration, Systems and Cosmology

In physics and other fields, the definition of duration is a fundamental problem. We are used to astronomical time based upon the idea that the rotation of Earth is perfectly regular, an assumption which nowadays we know to be slightly erroneous. A unit of time based upon the period of a chosen atomic vibration is preferred. In all cases we need a time which we accept as a standard, taken for granted. But there are cases where duration associated with such a universal time does not seem appropriate. At certain instants it seems to an individual that time elapses more slowly or more quickly. This psychological time is subjective; it depends upon the person concerned and the circumstances. For a given individual it also depends upon age; at the end of life a day seems shorter than in youth. Of course this "relativity" of duration has nothing to do with relativity of time met in special relativity and is not at all in opposition to it.

If we have chosen a standard or reference time t, such as for example the astronomical one, what is the most general time θ we can derive from it as a function θ(t)? We assume that θ(t) must be a continuous function of t (though a discrete time could be proposed) and add that θ(t) must not decrease when t increases. More precisely we may write, if f (a,b) is the duration of interval (a,b), and since f(a,b) must increase in the large with b and decrease in the large with a:

$$(a,b) + f(b,c) = f(a,c)$$
$$f(a,b) \geq 0, \; b \geq a$$
$$f(a,a) = 0.$$

If we add the hypothesis that f is differentiable, we have,

$$f(a+da,\ b)\ +f(b,\ c+dc)\ = f(a+da,\ c+dc).$$

Replacing $f(a+da,b)$ *by* $f(a,b)+\partial f(a,b)/\partial a$ da and the like for $f(b,c+dc)$ *and* $f(a+da,c+dc)$ we obtain,

$$\partial f(a,b)/\partial a\ =\ \partial f(a,c)/\partial a.$$

So $\partial f(a,b)/\partial a$ is independent of b. Consequently, after integration with respect to a, we have,

$$f(a,b)\ =\ F(a)\ +\ Cst,$$

the integration constant being a function of b or G(b). It gives,

$$f(a,b)\ =\ F(a)\ +\ G(b).$$

But, since f(a,a) = 0, we have G = - F and so,

$$f(a,b)\ =\ G(b)\ -\ G(a)$$

Or,

$$f(a,b)\ =\ \theta(b)\ -\ \theta(a),$$

Function θ (t) being obviously continuous at not decreasing with t, as required. We consider now a dynamical system. This involves in its evolution equation a reference time t which, in a way, is impartial. But since it has nothing to do with the considered system it does not take into account its intrinsic behaviour. Metaphorically all reference instants have not the same value. If the system were conscious, some instants, or short intervals of reference time, would have a greater importance than others. In the extreme case of very profound sleep or, better, of a coma, duration is not felt. This is close to the point of view of Aristotle in chapter IV of his "Physics": "When we feel no change in our thought, or we are unconscious of this change, or when we feel it without being aware of it, then it seems to us that no time have elapsed". Augustine in book XI of his "Confessions" expresses an opinion not far from that of Aristotle: "What is time? If nobody asks, I know; but if I want to explain, I do not know! Nevertheless – I tell it confidently – I know that if nothing happened, there would be no time passed" Finally we are inclined to propose as a first approach that the more rapidly the state of the system changes, the more important are the corresponding reference instants.

We choose, as an index of importance of reference instant t, the scalar square of the speed of evolution of the state at this instant, that is to say $\left(dX(t)\ \ dt\right)$. Of course many other indexes are possible, given for example by a strictly increasing function of the modulus of the speed. So we propose as an intrinsic or "internal duration" $d(t_1,t_2)$ of reference interval (t_1,t_2) the integral,

$$\int t_1,t_2\ \left(dX(t)/dt\right)^2 dt,$$

The internal duration of infinitesimal interval (t, t+dt) being $(dXt)\ \ dt\ \ dt$. An "internal time", coherent with this duration and defined up to an additive constant, is given by,

$$\theta(t) = d(t_0,t)$$

And we have,

$$d(t_1,t_2) = \theta(t_2) - \theta(t_1),$$

A result which is in accordance with what we expected from the most general time we can derive from a given standard or reference time t.

Explosions and Implosions

This internal time may be used for any dynamical system defined by a differential equation. We have particularly considered what we have called "elliptic explosion-implosion", "hyperbolic explosion" and, as an intermediary case, "parabolic explosion.

Elliptic Explosion-Implosion

In the case of an "elliptic explosion-implosion", the equation of evolution is given by,

$$dX(t)/dt = q/p\ sgn(p-t)\ \left(q^2 - X^2(t)\right)^{1/2} / X(t)$$

Where the state X(t) is a mere scalar with,

$$X(0) = 0,\ p > 0,\ q > 0,\ 0 \leq t \leq 2p.$$

It is easy to see that,

$$X(t) = q/p\ \left(p^2 - (p-t)^2\right)^{1/2}$$

Since by derivation it gives,

$$dX(t)/dt = q/p\ (p-t)\ /\ \left(p^2 - (p-t)^2\right)^{1/2}$$

which is the expression obtained from $dX(t)/dt = q/p\ sgn(p-t)\ \left(q^2 - X^2(t)\right)^{1/2} / X(t)$ if we replace X(t) by $X(0) = 0,\ p > 0,\ q > 0,\ 0 \leq t \leq 2p.$ The graph of function X(t), which represents the evolution of state X(t)with reference time t, is the upper part of an ellipse of great axis 2p and small axis q.The absciss of the centre is p and its ordinate is 0. X(t) starts from 0 at t=0, increases to its maximum value q at t= p, then decreases and attains 0 at t=2p. The speed at t= 0 is +∞ and -∞ at t= 2p. That is why we have an explosion at the beginning and an implosion at the end, and so what we can call an "elliptic explosion –implosion". The square of the speed of evolution is according to (3), after a very classical decomposition of,

$$q^2/p^2\ (t-p)^2 / \left(p^2 - (p-t)^2\right)\ 1/2 = q^2/p^2\ (p-t)^2 / t(2p-t),$$

Given by,

$$\left(d\left(X(t)\right)/dt\right)^2 = q^2/p^2\left(q^2 - X^2(t)\right)/X2(t) = q^2/2p\left(1/t - 2/p + 1/2p-t\right).$$

The "internal time" we can obtain by integration is defined up to an additive constant we can choose freely. The simplest choice gives,

$$\theta(t) = q^2/2p\ (Logt - 2t/p - Log(2p-t)$$

We see that when the reference time t varies from 0 to 2p, the "internal time" varies from $-\infty$ to $+\infty$. Obviously this circumstance, the push back of $t = 0\ to\ \theta = -\infty$ and the push forward of $t = 2p\ to\ +\infty$, is linked to the behaviour of the square of the speed near t = 0 and near t = 2p which generates logarithms.

Hyperbolic Explosion

In the case of "hyperbolic explosion" the equation of evolution is,

$$dX(t)/dt = q/p\ \left(q^2 + X^2(t)\right)/X(t)\ ,$$

With,

$$X(0) = 0\ , p > 0,\ q > 0,$$

$$0 \le t$$

It is easy to verify that we have,

$$X(t) = q/p\ \left((p+t)^2 - p^2\right)^{1/2},$$

Since by derivation,

$$dX(t)/dt = q/p\ (p+t)/\left((p+t)^2 - p^2\right),$$

Expression also obtained from $dX(t)/dt = q/p\ \left(q^2 + X^2(t)\right)/X(t)$, when we replace X(t) by $X(t) = q/p\ \left((p+t)^2 - p^2\right)^{1/2}$. The graph of function X(t) is the upper half right part of an hyperbola. The absciss of its centre is –p and its ordinate is 0. The asymptote associated with the graph of X(t) has a slope equal to q/p. X(t) starts from 0 at t = 0, then tends to $+\infty$ when t tends to $+\infty$. For great values of t, X(t) behaves as (q/p) t + q. The square of the speed of evolution is,

$$\left(dX(t)/dt\right)^2 = q^2/2p\left(1/t + 2/p - 1/2p + t\right).$$

This gives an "internal time" equal to,

$$\theta(t) = q^2/2p\ \left(Logt + 2t/p - Log(2p+t)\right)$$

For which when reference time t varies from 0 to $+\infty$, "internal time" varies from $-\infty$ to $+\infty$.

Parabolic Explosion

The "parabolic explosion" is an intermediary case, as parabola is "intermediary" between ellipse and hyperbola. Starting from equation $dX(t)/dt = q/p\, sgn(p-t)\, \left(q^2 - X^2(t)\right)^{1/2} / X(t)$, we shall make p tend to ∞ while keeping q2/p equal to a constant h. We have,

$$dX(t)/dt = (h/p)^{1/2}\left(hp - X2(t)\right)^{1/2} / X(t)$$

$$= \left(h^2 - h/p\, X^2(t)\right)^{1/2} / X(t),$$

which gives, when p tends to ∞, the new equation of evolution,

$$dX(t)/dt = h/X(t)$$

With,

$$X(0) = 0,$$

And,

$$h > 0.$$

So,

$$X(t) = (2ht)^{1/2}.$$

The graph of function X (t) is the upper part of a parabola of summit at t = 0 and having t axis as axis. It is the limit of the half ellipse seen in the elliptic case. We have an explosion at t = 0 with initial speed $+\infty$. This speed decreases with time and tends to 0 while X (t) tends to $+\infty$. The square of the speed is,

$$\left(dX(t)/dt\right)^2 = h/2t,$$

Giving the "internal time",

$$\theta(t) = h/2\, Logt,$$

which varies from $-\infty$ to $+\infty$ when t varies from 0 to $+\infty$.

Infinite Internal Duration

In the three cases seen above we have observed the possibility of an infinite "internal duration" linked to the push back (or forward) of a particular reference instant. Obviously this is linked to the behaviour of $(dX(t)/dt)^2$ near this reference instant. We choose, to simplify the presentation, reference instant t = 0, and suppose that X(t) is an analytic function near this point. So X(t) behaves near t = 0 as t n, dX(t)/dt as t n-1, (dX(t)/dt) 2 as t 2n-2 and so ∫ $(dX(t)/dt)^2$ dt as $t^{\frac{2n-1}{2n-1}}$. If n is different from 1/2, there is no singularity and no push back of t = 0 to $\theta_2 = -\infty$. But if n = ½ X(t) behaves as t 1/2, dX(t)/dt as t -1/2, $(dX(t)/dt)^2$ as t -1 and $\int (d(X(t)/dt))^2$ dt as Log t. There is a push back of t = 0 to $\theta = -\infty$ and possibility of infinite interval time. A push forward of reference instant t = a to $\theta = +\infty$ happens if X(t) behaves as (a-t) 1/2 near t = a. In the case of "elliptic explosion-implosion", X(t) behaves as $t^{1/2}$ near t = 0 and as (2p-t) 1/2 near t = 2p. So, as we have seen, we have a push back and a push forward. For "hyperbolic explosion" as well as for "parabolic explosion" X(t) behaves as $t^{1/2}$ near t = 0 and there is a push back.

Equation of Evolution in Term of "Internal Time"

It may be of interest, for a given evolution of a system described by function X(t), to express X(t) in term of "internal time" θ instead of reference time t. Let us take as an example the case of "parabolic explosion". We have,

$$X(t) = (2ht)^{1/2}$$

and,

$$\theta(t) = h/2 \text{ Log} t,$$

$$t(\theta) = \exp(2\theta/h)$$

It gives,

$$X(t(\theta)) = (2h)^{1/2} \exp(\theta/h).$$

So, in term of "internal time", the state varies exponentially from $0 \text{ to } +\infty$ while θ varies from $-\infty \text{ to } +\infty$, instead of growing as t ½ when t goes from $0 \text{ to } +\infty$.

Time and Space

The system considered may, more generally, be defined by X(t,x), a scalar function of "reference time" t and space point x, satisfying a partial derivative equation. We consider, as the index of importance of reference instant t, the integral, supposed to be convergent, extended to whole space S, of the square of the speed of evolution $(\partial X(t,x)/\partial t)^2$, that is to say,

$$\int_s (\partial X(t,x)/\partial t)^2 \, dx.$$

So the "internal duration" of interval (t_1, t_2) is given by,

$$d(t_1,t_2) = \int t_1,t_2 \int_s (\partial X(t,x)/\partial t)^2 \, dx \, dt,$$

and, an "internal time" by,

$$\theta(t) = d(t_0,t).$$

We shall apply this formalism to the dynamical system constituted by a space-time field of temperatures, in the case of heat diffusion, with $S = (-\infty,+\infty)$. Temperature at point x, at reference instant t, is u(t,x). The partial derivative equation of evolution is,

$$\partial u(t,x)/\partial t - \partial^2 u(t,x)/\partial x^2 = 0.$$

If the repartition of temperatures at t = 0 is given by function (more generally distribution) $u_0(x)$, the solution of the above equation is,

$$u(x,t) = \int_{-\infty\ +\infty} 1/2(\pi t)^{1/2}\ exp\left(-(x-s)^2/4t\right)\ u_0(s)\ ds.$$

At initial reference instant t = 0, we suppose that the field of temperatures is given by $\delta(x)$ or Dirac distribution centred at x= 0 (in a rather simplified language it is equal to 0 everywhere except at t = 0 where it is infinite, the integral being nevertheless equal to 1). The repartition of temperatures at reference instant t is, according to the precedent equation and the properties of $\delta(x)$, given classically by the Laplace-Gauss function,

$$u(t,x) = (4\pi t)^{1/2}\ exp\left(-x^2/4t\right).$$

When reference instant t tends to +∞, this function "flattens" and tends to $\varepsilon(x)$ or "epsilon distribution", in short it is equal to zero everywhere, the integral being nevertheless equal to 1. We have,

$$\partial u(t,x)/\partial t)^2 = 1/16\pi\left(1+x^2/2t\right)^2/\ t^3\ exp\left(-x^2/2t\right)$$

and,

$$\int_R \left(\partial u(t,x)/\partial t\right)^2\ dx = 3\ (2\pi)^{1/2}/16\ t^{-5/2}.$$

So the "internal duration" of reference interval (t_1, t_2) is, by integration from t_1 to t_2,

$$d(t_1,t_2) = (2\pi)^{1/2}/8\ \left(t_1^{-3/2} - t_2^{-3/2}\right)$$

and, an "internal time" is given by the following function (increasing with t),

$$\theta(t) = -\ (2\pi)^{1/2}/8\ t^{-3/2}.$$

When reference time t varies from $0\ to\ +\infty$, "internal time" θ varies from $-\infty\ to\ 0$. Initial reference instant t = 0 is pushed back to $-\infty$. "Internal duration" from $t>0\ to\ +\infty$, is finite and equal to $(2\pi)^{1/2}/8\ t^{-3/2}$.

Time and Cosmology

We shall now interpret the notion of "internal time" in the field of cosmology. We consider models for which the state of the universe, at reference instant t, is given by the so called scalar factor R(t). According to Lemaître, Friedman and Robertson a possible equation of evolution is,

$$\left(dR(t)/dt\right)^2 = 8\pi G/3\ \rho(t)\ R^2(t) - kc^2\ +\ \Lambda/3\ R^2(t),$$

R (o) = o,

G being the gravitational constant, c the speed of light, k the index of curvature (k= -1, space with negative curvature; k = o, flat space; k =+1, space with positive curvature), Λ the cosmological constant, $\rho(t)$ the density of matter equal to a/R³ (t)) or its material equivalent b/R⁴ (t) when there is only radiation, a and b being two constants. When k = +1, R (t) is interpreted as the radius of the universe.

Radiation with Null Cosmological Constant

If we consider the case of positive curvature with null cosmological constant and density of matter negligible compared to the equivalent density of matter of pure radiation (k=+1, Λ=o, $\rho(t)$ = b/ R^4(t)), we have,

$$\left(dR(t)/dt\right)^2 = 8\pi G/3\ b/R^2(t) - c^2$$

and,

R (o) = o.

This case corresponds to the "elliptic explosion-implosion" considered above where, after having taken the square of the two members of equation $dX(t)/dt = q/p\ sgn(p-t)\ \left(q^2 - X^2(t)\right)^{1/2} / X(t)$, we replace X (t) by R(t), choose, q = cp,

and,

$$p = \left(b8\pi G/3\right)^{1/2}/c^2.$$

It gives, according to $X(0) = 0,\ p > 0,\ q > 0,\ 0 \le t \le 2p.$,

$$R(t) = q/p(2pt - t\ 2)^{1/2} = \left(2\left(8\pi Gb/3\right)^{1/2} t - c^2t^2\right)^{1/2}$$

and the graph of function R(t) is, as we know, elliptical.

The "internal time "of this cosmological system is, according to equation,

$$dX(t)/dt = q/p\ (p-t)\ /\ \left(p^2 - (p-t)\ ^2\right)^{1/2}$$

Or,

$$\theta(t) = \left(2\pi Gb/3\right)^{1/2}(Logt - tc^2/\left(2\pi Gb/3\right)^{1/2} - Log\left(2\left(2\pi Gb/3\right)^{1/2} - t\right).$$

While reference time t goes from o (big bang) to t = 2p (big crunch) "internal time" θ goes from $-\infty$ *to* $+\infty$. In that case I propose to call θ "generalized cosmological time" in remembrance of "cosmological time" given by,

$$c^2p/2\ Logt$$

Or,

$$(2\pi Gb/3)^{1/2}\ Logt.$$

which is approximately valid for t "small".

If we consider now the case of flat space, null cosmological constant and pure radiation,

k = 0, Λ = 0,

$$\rho(t) = b/R^4(t)),$$

We have,

$$(dR(t)/dt)^2 = 8\Pi G/3\ b/R^2(t),$$

Or,

$$dR(t)/dt = 2\ (2\Pi Gb/3)^{1/2}/R(t).$$

We recognize, according to $X(t)=(2ht)^{1/2}$, a "parabolic explosion" with $h=2(b2\Pi G/3)^{1/2})$. We have,

$$R(t) = 2\ (b2\Pi G/3)^{1/4}\ t^{1/2}$$

and, according to $\theta(t)=h/2\ Logt$,, the "internal time" is given by,

$$\theta(t) = (2\pi Gb/3)^{1/2}\ Logt,$$

identical to the approximate formula $(dR(t)/dt)^2=8\Pi G/3\ \rho(t)\ R^2(t)-kc^2\ +\ \Lambda/3\ R^2(t)$. While t goes from 0 (big bang) to + ∞, θ varies from – ∞ to + ∞.

Neither Matter nor Radiation

There are other cases for which we can introduce "internal time". Some of them may not be realistic, but due to the uncertainty concerning our conception of the universe and its evolution, they must not be discarded systematically. For example we may have a universe with no matter and no radiation, at least as an approximation. As a first case we add that space has a negative curvature and a negative cosmological constant,

k = -1, Λ < 0,

$$\rho(t) = 0.$$

We have,

$$(dR(t)/dt)^2 = c^2\ +\ \Lambda/3\ R^2(t),\ R(0) = 0,$$

$$dR(t)/dt = (c^2\ +\ \Lambda/3\ R^2(t))^{1/2},$$

which gives,

$$R(t) = c\ (3/|-\Lambda|)^{-1/2}\ sin\left(t\ \left(|-\Lambda|/3\right)^{1/2}\right).$$

R(t) starts from 0 (big bang) reaches its maximum $c(3/|-\Lambda|)^{1/2}$,decreases and attains 0 at t = 2⊣ $(|-\Lambda|/3)^{-1/2}$ (big crunch). We have,

$$\left(dR(t)/dt\right)^2 = c^2\ cos^2\left(t\ \left(|-\Lambda|/3\right)^{1/2}\right) = c^2/2\ \left(1+cos\ 2t\ \left(|-\Lambda|/3\right)^{1/2}\right)$$

which give after integration,

$$\theta(t) = c^2/2\ \left(t\ +\ sin2t\left(|-\Lambda|/3\right)^{1/2}\ \right)/2\left(|-\Lambda|/3\right)^{1/2}.$$

So θ varies from 0 to $c2\pi/4\ (|-\Lambda|/3)^{-1/2}$ as t varies from 0 to $\pi/2\ (|-\Lambda|/3)^{-1/2}$. A finite reference duration gives here a finite "internal duration".

Another possibility is the case of universe, with no matter nor radiation as above but with positive curvature and positive cosmological constant,

$k = +1, \Lambda > 0,$

$$\rho(t) = 0.$$

We have,

$$\left(dR(t)/dt\right)^2 = -c^2 + \Lambda/3\ R^2(t),$$

$$dR(t)/dt = \left(-c^2\ +\ \Lambda/3\ R^2(t)\right)\ 1/2,$$

and so,

$$R(t) = c\ (3/\Lambda)^{1/2}\ cosh\left(t\left(\Lambda/3\right)^{1/2}\right).$$

R(t) starts from $c(3/\Lambda)\ 1/2$ at t= 0 and tends, in a way closer and closer to an exponential to $+\infty$ as t tends to $+\infty$. We have,

$$\left(dR(t)/dt\right)^2 = c^2\ sinh^2\left(t\left(\Lambda/3\right)^{1/2}\right),$$

and after integration,

$$\theta(t) = c^2/2\ (-t\ +\ cosh2\left(t\left(\Lambda/3\right)^{1/2}/2\left(\Lambda/3\right)^{1/2}\right),$$

Which vary from $c^2/2$ to $+\infty$ when t varies from 0 to $+\infty$ (not forgetting that θ is defined up to an arbitrary constant). Infinite reference duration gives an infinite "internal duration".

Matter but no Radiation

We shall consider some other cases, also presented by Berry, where radiation is negligible. We start with the hypothesis of a flat space and a negative cosmological constant,

$$k=0, \Lambda$$

$$\rho(t) = a / R^3(t).$$

We have,

$$\left(dR(t)/dt\right)^2 = 8\Pi Ga/3/R(t) + \Lambda/3\, R^2(t),$$

Which gives, A being a constant,

$$R(t) = A\, sin^{2/3}\left(t/2\, (3|\Lambda|)^{1/2}\right).$$

So R (t) starts from 0 at t = 0 with infinite speed (big bang) reaches its maximum, decreases and attains 0 again for $t = 2\pi/(3|\Lambda|)^{1/2}$ (big crunch). Near t = 0 R (t) behaves as $t^{2/3}$ which is different from $t^{1/2}$. So, as we have seen, there is no push back of reference instant t= 0 to -∞ and, for analogous reasons, no push forward of instant $t = 2\pi/(3|\Lambda|)^{1/2}$ *to* $+\infty$.

We have now the intermediary case where the cosmological constant is equal to zero (k=0, Λ=0, $\rho(t) = a/R^3(t)$). which gives,

$$\left(dR(t)/dt\right)^2 = 8\Pi Ga/3\ /R(t)$$

and,

$$R^{1/2}(t)\ dR(t)/dt = \left(8\Pi Ga/3\right)^{1/2}$$

Or,

$$R(t) = \left(8\Pi Ga/3\right)^{1/3}\ t^{2/3}$$

When t varies from 0 to +∞, R(t) increases from 0 with infinite speed (big bang) to +∞.

We have,

$$\left(dR(t)/dt\right)^2 = 4/3\left(8\Pi Ga/3\right)^{2/3}\ t^{-2/3},$$

which gives after integration,

$$\theta(t) = 3\left(8\Pi Ga/3\right)^{2/3}\ (2/3)^{2/3}\ t^{1/3}.$$

There is no push back of reference instant t = 0.

Now we must see the case where the cosmological constant is positive (k=0, Λ>0, $\rho(t) = a / R$). We have,

$$\left(dR(t)/dt\right)^2 = 8\pi G/3\, a/R(t) + \Lambda/3\, R^2(t)$$

and B being a constant,

$$R(t) = B\, sinh^{2/3}\left(t/2\, (3\Lambda)^{1/2}\right).$$

R (t) starts from zero at t=0 with infinite speed (big bang) and tends to $+\infty$ exponentially. There is no push back of reference instant t=0 to $-\infty$.

Null Cosmological Constant and no Radiation

There are other interesting cases with negligible radiation and null cosmological constant. We start with a space of negative curvature ($k=-1, \Lambda=0, \rho(t) = a/R^3(t)$). We have,

$$\left(dR(t)/dt\right)^2 = 8\pi Ga/3\ /R(t) + c^2.$$

It is easier to represent the graph of function R(t) parametrically than explicitly. This gives with 0 ≤ u,

$$R(t) = 8\pi Ga/6\, (coshu - 1),$$
$$t = 8\pi Ga/6c\, (sinhu - u).$$

R (t) starts from 0 at reference instant t= 0 with infinite speed (big bang), then tends to $+\infty$ asymptotically as ct. Near t =0, R(t) behaves as t $^{1/3}$. It proves as we have already seen that there is no push back of reference instant t =0 to -∞. If space is flat ($k=0,\ \Lambda=0, \rho(t) = a/R3$) we find a case already studied above, we have,

$$\left(dR(t)/dt\right)^2 = 8\pi Ga/3/R(t)$$

R(t) starts from 0 at reference instant t= 0 with infinite speed (big bang), then tends to +∞ asymptotically as ct. Near t =0, R(t) behaves as t 1/3. It proves as we have already seen that there is no push back of reference instant t =0 to -∞. If space is flat ($k=0,\ \Lambda=0, \rho(t) = a/R3$) we find a case already studied above, we have,

$$\left(dR(t)/dt\right)^2 = 8\pi Ga/3/R(t) - c^2.$$

The graph of function R(t) is a cycloid represented parametrically by,

$$R(t) = 8\pi Ga/6\, (1 - cosv),$$

$$t = 8\pi Ga/6c\, (v - sinv),$$

$$0 \le v \le \pi.$$

R(t) starts from 0, at reference instant t=0, with an infinite speed (big bang). It increases up to 8π Ga/6 attained at t = $8\pi Ga / 6c$ ($\pi/2$ −1), then decreases to 0 attained at t = $8\pi Ga / 6$ (π−2). Near t = 0, R(t) behaves as t 2/3 and so there is no push back of instant t = 0 to $-\infty$, and for similar reasons no push forward of instant t= $8\pi Ga / 6$ (π−2) to $+\infty$.

Another Approach to "Internal Time"

This new approach will put, metaphorically speaking, emphasis on perception. The purpose being is to propose a modelling of the perception duration. First we consider the linear differential equation,

$$dx(t)/dt = -a(t)\, x(t) + v(t),$$

where t is reference time, x(t) and v(t) two scalar functions. We have classically,

$$x(t) = \varphi(t, t_0)\, x(t_0) + \int_{t_0,t} \varphi(t,\tau)\, v(\tau)\, d\tau$$

where,

$$\varphi(t, t_0) = exp\left(- \int_{t0},\, t\, a(s)\, ds\right)$$

with the hypothesis that $\varphi(t, t_0)$ tends to 0 if t tends to $+\infty$. When a(t) is a mere constant a, it means obviously that a is strictly positive. The sign – has been placed before the integral to make more evident that the positivity of this consequence. Let us remark that we have the following property of "transitivity",

$$\varphi(t',t)\, \varphi(t,\tau) = \varphi(t',\tau).$$

We interpret v(t) as an external influence. In the simplest case we have,

$$v(t) = b(t)\, u(t),$$

more generally we could have,

$$v(t) = b_0(t)\, u(t) + b_1(t)\, du(t)/dt + \ldots$$

Involving derivatives of u(t). This formulation is nor irrealistic and is well adapted to the modelization of a tachymeter or an accelerometer if we consider mainly the first or the second derivatives the other terms being rather negligible. This formula which we may also write with the help of the Dirac distribution τ and its derivatives,

$$v(t) = \int_{-\infty,+\infty} \left(b_0(\tau)\, \vdash(t-\tau) + b_1(\tau)\, \tau'(t-\tau) + \ldots\right) u(\tau)\, d\tau$$

Each b_i ((t) is a "factor of attention" concerning a particular derivative. The passage of function u to function v is made by what we call "observation operator". Here this operator acts in an instantaneous way, being purely local. The first factor b_0 (t), or more simply b(t), may be considered positive (when it is null there is no attention and so no perception at the considered instant t). This factor of attention has been pointed out: "It remains an impression more or less strong according to the fact that the attention has been more or less intense". More generally the passage of u to v through v = O(x), O being an "observation operator", is not instantaneous but hereditary, that is to say involving the past and present of u. This has been observed: "In fact 'pure' perception, that is to say instantaneous, is only an ideal, a limit. Every perception fills a certain length of duration, extends the past in the present". For example we may have a convolution,

$$v(t) = \int_{t_0,t} k(t-\tau)\, u(\tau)\, d\tau,$$

where v(t) depends upon the values of u on interval (t_0,t). More generally, if we do not leave the case of linear "observation operators" we have a Volterra composition giving,

$$v(t) = \int t_0\, t\, k(t,\tau)\, d\tau\ .$$

The formalism of "observation operators" permits to see in which cases such an operator does not alter the observed function u. We must have,

$$O(u) = \lambda\, u,$$

So u must be an Eigen function of operator O and λ is the associated Eigen value which may be complex. If λ = 1, we have a fixed point. We give these details about "observation operators" because we shall meet them in several circumstances, the problem of appreciation of time having to do with observation and more generally with what we could call "mathematical epistemology".

Let us come back to the differential equation and its solution. We interpret $\varphi(t,\tau)\, v(\tau)$ as what remains at instant t of the perception $v(\tau)$ felt at anterior instant τ, it is the result of the transfer by memorization of $v(\tau)$ from τ to t. The property of transitivity $\varphi(t,\, t_0) = exp\left(-\int_{t0},\, t\, a(s)\, ds\right)$ makes this transfer coherent. According to the hypothesis that $\varphi(t,\tau)$ tends to 0 if $t-\tau$ tends to $+\infty$, we may say that the transferred perception $\varphi(t,\tau)v(\tau)$ tends to 0 if τ tends to $-\infty$. In other words we may conclude that the more ancient is a perception the more feeble is its remembrance. If our differential system starts with the null state we have according to $x(t) = \varphi(t,\, t_0)\, x(t_0) + \int_{t_0,t} \varphi(t,\tau)\, v(\tau)\, d\tau$ a Volterra composition which reduces to convolution when a(s) is a constant a,

$$x(t) = \int_{t0,\,t} \varphi(t,\tau)\, v(\tau)\, d_\tau.$$

We may interpret x(t) as the result of the superposition of all the successive perceptions, from to to t, as they are transferred to t by memorization. The passage from v(t) to x(t) is given by a special type of linear "observation operator" which we call "memorization operator". It may be compared to the factor of forgetfulness or the memory coefficient.

We shall consider now differential equation $dx(t)/dt = -a(t)\, x(t) + v(t)$, or more precisely its solution $x(t) = \varphi(t,\, t_0)\, x(t_0) + \int_{t_0,t} \varphi(t,\tau)\, v(\tau)\, d\tau$ where we replace x(t) by θ(t), as a model of perception and memorization of duration valid for a dynamical system considering t as "reference time" and θ(t) as a subjective or "internal time", even if these expressions are to be understood metaphorically. Taking into account $\varphi(t,\, t_0) = exp\left(-\int_{t0},\, t\, a(s)\, ds\right)$ we have,

$$\theta(t) = \int_{t_0,t} exp\left(-\int \tau,\, t\, a(s) ds\right) v(\tau)\, d\tau,$$

to which we add,

$$a(t) \geq 0, v(t) \geq 0.$$

While t – t o is the reference duration of interval (t_0, t), $\theta(t) - \theta(t_0) = \theta(t)$ is its subjective or internal duration. It depends upon the way the considered system perceives through v(t) and memorizes, not upon the way the state of the system evolves as it was the case in the first approach to "internal time".

Of course t is a good parameterization of time, in the sense that two distinct instants are represented by two different values of t. If θ(t) is to be a good parameterization of time, it must be a strictly increasing function of t, a condition realized if v(t) never vanishes on an interval not reduced to a mere instant. The hypothesis that a(t) ≥ o has for consequence that the factor of memorization $\varphi(t, \tau)$ decreases in the large when τ diminishes. We must now interpret function v(t) in the most simple case where,

$$v(t) = b(t)\, u(t),$$

With,

$$b(t) \geq 0\ , u(t) \geq 0.$$

Of course more elaborated cases could be considered, making a full use of "observation operators".

We consider that function u(t) gives the evolution of the weight of importance of reference instant t. The very simple "observation operator" represented by b(t) appears as a "factor of attention" given to u(t) which represents the intrinsic importance of reference instant t itself. Since t is by definition an objective time or "reference time", all instants t have an equal intrinsic importance which we may decide to be equal to 1. So we may write simply,

$$vt) = b(t)$$

As the weight attributed to instant t is equal to the "factor of attention" b(t) at this instant.

Perception of Duration with Imperfect Memorization

In our model we have an imperfect memorization when a(t) is not identical to o. The factor of memorization,

$$\varphi(t,\tau) = exp\left(- \int_{\tau,t} a(s)ds\right)$$

is strictly inferior to 1 and tends to o when t – τ tends to + ∞, so when τ tends to - ∞. The remembrance of past perceptions vanishes with time. Since, as we have seen, v(t) is equal to b(t) we may write according to $\varphi(t',t)\ \varphi(t,\tau) = \varphi(t',\tau)$,

$$\varphi(t,\tau)=\int_{t_0},\ t\ exp\left(-\int\ \tau,\ t\ a(s)ds\right)\ b(\tau)d\tau,$$

Or,

$$d\theta(t) = \left(-a(t)\ \theta(t) + b(t)\right)\ dt,$$

Explaining the antagonistic roles of b(t) > o and -a(t) θ(t) < o, representing attention on one side and oblivion on the other. Or, if we consider the case where a(t) reduces to constant a>o,

$$\theta(t) = \exp(-at)\int to, t\ exp\ a\tau\ b(\tau)d\tau.$$

It is interesting to see what happens when b(t) is a Dirac distribution $\delta(t-\gamma)$, centred on instant γ. It is an idealisation which gives,

$$\theta(t) = exp(-a(t-\gamma)),$$

$$t > \gamma.$$

We see that the reference duration of instant γ, obviously equal to o, is perceived just after this instant, due to the infinite attention implied, as having a finite subjective or internal duration; It is perceived later has having a decreasing value tending to o. If we replace the Dirac γ by a flash of attention, things are not so sharply defined but are of the same nature: a very short interval of reference time is perceived as middle sized interval of "internal time" and this impression diminishes with time and disappears. The reference time, 1/a, necessary to see the perceived duration divided by e is a measure of what we may call the "subjective duration of the present" or the "thickness of an instant", in accordance with Bergson's remark quoted above and close to the concept of constant of time familiar in dynamics.

$$d\theta(t)/dt = b(t),$$

With,

$$b(t) \geq 0,\ \theta(t0) = 0.$$

We have,

$$\theta(t) = \int t_0, t\ b(\tau)\ d\tau,$$

it is the subjective or "internal duration" of interval (t_0,t). In other terms it is the subjective duration of instants of consciousness in this interval, instants where b(t) = o having no impact (metaphorically or not : deep sleep or coma). If b(t) takes only values 1 or o, θ(t) is equal to the internal duration (as well as reference duration) of instants of consciousness. When b(t) takes value 1 at discrete instants it acts as a stroboscopic selecting device. When b(t) is constant θ(t) is proportional to $t-t_0$. If for example, the "factor of attention" decreases exponentially with time that is to say if,

$$b(t) = b\ exp(-\lambda t),$$

With,

$$b > 0,\ \lambda > 0.$$

We have, with t_0 = o,

$$(\theta t) = 1 - exp(-\lambda t).$$

The “internal duration” or perceived duration since $t_0 = 0$, increases from 0 to 1 while “reference duration” tends to $+\infty$. This circumstance may be found during the observation of a disintegrating radioactive material if the factor of attention b(t) is proportional to disintegration activity which decreases exponentially.

Big Bang

How was our Universe created? How did it come to be the seemingly infinite place we know of today? And what will become of it, ages from now? These are the questions that have been puzzling philosophers and scholars since the beginning the time, and led to some pretty wild and interesting theories. Today, the consensus among scientists, astronomers and cosmologists is that the Universe as we know it was created in a massive explosion that not only created the majority of matter, but the physical laws that govern our ever-expanding cosmos.

This is known as The Big Bang Theory. For almost a century, the term has been bandied about by scholars and non-scholars alike. This should come as no surprise, seeing as how it is the most accepted theory of our origins. But what exactly does it mean? How was our Universe conceived in a massive explosion, what proof is there of this, and what does the theory say about the long-term projections for our Universe?

The basics of the theory are fairly simple. In short, the Big Bang hypothesis states that all of the current and past matter in the Universe came into existence at the same time, roughly 13.8 billion years ago. At this time, all matter was compacted into a very small ball with infinite density and intense heat called a Singularity. Suddenly, the Singularity began expanding and the universe as we know it began.

While this is not the only modern theory of how the Universe came into being – for example, there is the Steady State Theory or the Oscillating Universe Theory – it is the most widely accepted and popular. Not only does the model explain the origin of all known matter, the laws of physics, and the large scale structure of the Universe, it also accounts for the expansion of the Universe and a broad range of other phenomena.

Timeline

Working backwards from the current state of the Universe, scientists have theorized that it must have originated at a single point of infinite density and finite time that began to expand. After the initial expansion, the theory maintains that Universe cooled sufficiently to allow the formation of subatomic particles, and later simple atoms. Giant clouds of these primordial elements later coalesced through gravity to form stars and galaxies.

This all began roughly 13.8 billion years ago, and is thus considered to be the age of the universe. Through the testing of theoretical principles, experiments involving particle accelerators and high-energy states, and astronomical studies that have observed the deep universe, scientists have constructed a timeline of events that began with the Big Bang and have led to the current state of cosmic evolution.

However, the earliest times of the Universe – lasting from approximately 10-43 to 10-11 seconds after the Big Bang – are the subject of extensive speculation. Given that the laws of physics as we know them could not have existed at this time, it is difficult to fathom how the Universe could have been governed. What's more, experiments that can create the kinds of energies involved have not yet been conducted. Still, many theories prevail as to what took place in this initial instant in time, many of which are compatible.

Singularity

Also known as the Planck Epoch (or Planck Era), this was the earliest known period of the Universe. At this time, all matter was condensed on a single point of infinite density and extreme heat. During this period, it is believed that the quantum effects of gravity dominated physical interactions and that no other physical forces were of equal strength to gravitation.

This Planck period of time extends from point 0 to approximately 10^{-43} seconds, and is so named because it can only be measured in Planck time. Due to the extreme heat and density of matter, the state of the universe was highly unstable. It thus began to expand and cool, leading to the manifestation of the fundamental forces of physics.

From approximately 10^{-43} second and 10^{-36}, the universe began to cross transition temperatures. It is here that the fundamental forces that govern the Universe are believed to have began separating from each other. The first step in this was the force of gravitation separating from gauge forces, which account for strong and weak nuclear forces and electromagnetism.

Then, from 10^{-36} to 10^{-32} seconds after the Big Bang, the temperature of the universe was low enough (10^{28} K) that the forces of electromagnetism (strong force) and weak nuclear forces (weak interaction) were able to separate as well, forming two distinct forces.

Inflation Epoch

With the creation of the first fundamental forces of the universe, the Inflation Epoch began, lasting from 10^{-32} seconds in Planck time to an unknown point. Most cosmological models suggest that the Universe at this point was filled homogeneously with a high-energy density, and that the incredibly high temperatures and pressure gave rise to rapid expansion and cooling.

This began at 10^{-37} seconds, where the phase transition that caused for the separation of forces also led to a period where the universe grew exponentially. It was also at this point in time that baryogenesis occurred, which refers to a hypothetical event where temperatures were so high that the random motions of particles occurred at relativistic speeds.

As a result of this, particle–antiparticle pairs of all kinds were being continuously created and destroyed in collisions, which is believed to have led to the predominance of matter over antimatter in the present universe. After inflation stopped, the universe consisted of a quark–gluon plasma, as well as all other elementary particles. From this point onward, the Universe began to cool and matter coalesced and formed.

Cooling Epoch

As the universe continued to decrease in density and temperature, the energy of each particle began to decrease and phase transitions continued until the fundamental forces of physics and elementary particles changed into their present form. Since particle energies would have dropped to values that can be obtained by particle physics experiments, this period onward is subject to less speculation.

For example, scientists believe that about 10^{-11} seconds after the Big Bang, particle energies dropped considerably. At about 10-6 seconds, quarks and gluons combined to form baryons such as protons and neutrons, and a small excess of quarks over antiquarks led to a small excess of baryons over antibaryons.

Since temperatures were not high enough to create new proton-antiproton pairs (or neutron-antineutron pairs), mass annihilation immediately followed, leaving just one in 10^{10} of the original protons and neutrons and none of their antiparticles. A similar process happened at about 1 second after the Big Bang for electrons and positrons. After these annihilations, the remaining protons, neutrons and electrons were no longer moving relativistically and the energy density of the universe was dominated by photons – and to a lesser extent, neutrinos.

A few minutes into the expansion, the period known as Big Bang nucleosynthesis also began. Thanks to temperatures dropping to 1 billion kelvin and the energy densities dropping to about the equivalent of air, neutrons and protons began to combine to form the universe's first deuterium (a stable isotope of Hydrogen) and helium atoms. However, most of the Universe's protons remained uncombined as hydrogen nuclei.

After about 379,000 years, electrons combined with these nuclei to form atoms (again, mostly hydrogen), while the radiation decoupled from matter and continued to expand through space, largely unimpeded. This radiation is now known to be what constitutes the Cosmic Microwave Background (CMB), which today is the oldest light in the Universe.

As the CMB expanded, it gradually lost density and energy, and is currently estimated to have a temperature of 2.7260 ± 0.0013 K (-270.424 °C/ -454.763 °F) and an energy density of 0.25 eV/ cm^3 (or 4.005×10^{-14} J/m^3; 400–500 photons/cm^3). The CMB can be seen in all directions at a distance of roughly 13.8 billion light years, but estimates of its actual distance place it at about 46 billion light years from the center of the Universe.

Structure Epoch

Over the course of the several billion years that followed, the slightly denser regions of the almost uniformly distributed matter of the Universe began to become gravitationally attracted to each other. They therefore grew even denser, forming gas clouds, stars, galaxies, and the other astronomical structures that we regularly observe today.

This is what is known as the Structure Epoch, since it was during this time that the modern Universe began to take shape. This consists of visible matter distributed in structures of various sizes, ranging from stars and planets to galaxies, galaxy clusters, and super clusters – where matter is concentrated – that are separated by enormous gulfs containing few galaxies.

The details of this process depend on the amount and type of matter in the universe, with cold dark matter, warm dark matter, hot dark matter, and baryonic matter being the four suggested types.

However, the Lambda-Cold Dark Matter model (Lambda-CDM), in which the dark matter particles moved slowly compared to the speed of light is the considered to be the standard model of Big Bang cosmology, as it best fits the available data.

In this model, cold dark matter is estimated to make up about 23% of the matter/energy of the universe, while baryonic matter makes up about 4.6%. The Lambda refers to the Cosmological Constant, a theory originally proposed by Albert Einstein that attempted to show that the balance of mass-energy in the universe was static. In this case, it is associated with Dark Energy, which served to accelerate the expansion of the universe and keep its large-scale structure largely uniform.

Long-Term Predictions

Hypothesizing that the Universe had a starting point naturally gives rise to questions about a possible end point. If the Universe began as a tiny point of infinite density that started to expand, does that mean it will continue to expand indefinitely? Or will it one day run out of expansive force, and begin retreating inward until all matter crunches back into a tiny ball?

Answering this question has been a major focus of cosmologists ever since the debate about which model of the Universe was the correct one began. With the acceptance of the Big Bang Theory, but prior to the observation of Dark Energy in the 1990s, cosmologists had come to agree on two scenarios as being the most likely outcomes for our Universe.

In the first, commonly known as the "Big Crunch" scenario, the universe will reach a maximum size and then begin to collapse in on itself. This will only be possible if the mass density of the Universe is greater than the critical density. In other words, as long as the density of matter remains at or above a certain value ($1\text{-}3 \times 10^{-26}$ kg of matter per m^3), the Universe will eventually contract.

Alternatively, if the density in the universe were equal to or below the critical density, the expansion would slow down but never stop. In this scenario, known as the "Big Freeze", the Universe would go on until star formation eventually ceased with the consumption of all the interstellar gas in each galaxy. Meanwhile, all existing stars would burn out and become white dwarfs, neutron stars, and black holes.

Very gradually, collisions between these black holes would result in mass accumulating into larger and larger black holes. The average temperature of the universe would approach absolute zero, and black holes would evaporate after emitting the last of their Hawking radiation. Finally, the entropy of the universe would increase to the point where no organized form of energy could be extracted from it (a scenarios known as "heat death").

Modern observations, which include the existence of Dark Energy and its influence on cosmic expansion, have led to the conclusion that more and more of the currently visible universe will pass beyond our event horizon (i.e. the CMB, the edge of what we can see) and become invisible to us. The eventual result of this is not currently known, but "heat death" is considered a likely end point in this scenario too.

Other explanations of dark energy, called phantom energy theories, suggest that ultimately galaxy clusters, stars, planets, atoms, nuclei, and matter itself will be torn apart by the ever-increasing expansion. This scenario is known as the "Big Rip", in which the expansion of the Universe itself will eventually be its undoing.

Cosmological Spacetime

Standard cosmology is based on the Robertson-Walker (RW) metric for a spatially homogeneous and isotropic three-dimensional space, expanding or contracting as a function of time,

$$ds^2 = c^2dt^2 - a^2(t)\left[dr^2\left(1-kr^2\right)^{-1} + r^2\left(d\theta^2 \sin^2\theta d\phi^2\right)\right].$$

In the coordinates used for this metric, t is the cosmic time, measured by a commoving observer (and is the same everywhere), a(t) is the expansion factor, and r is an appropriately scaled radial coordinate in the co-moving frame. The geometric factor k is +1 for a closed universe, 0 for a flat universe, and −1 for an open universe.

The expansion of the universe is calculated from the Friedmann-Robertson- Walker (FRW) differential equations of motion,

$$H^2 \equiv \left(\frac{\dot{a}}{a}\right)^2 = \frac{8\pi G}{3c^2}\rho - \frac{kc^2}{a^2},$$

$$\frac{\ddot{a}}{a} = -\frac{4\pi G}{3c^2}(\rho + 3p),$$

$$\dot{\rho} = -3H(\rho + p),$$

derived from an application of the RW metric to the Einstein field equations. Here, an overdot denotes a derivative with respect to t, and ρ and p represent, respectively, the total energy density and total pressure. Often, the latter is written as p = w ρ, and w is then used to characterize the expansion properties of the medium. For example, w ≈ 0 for (visible and dark) matter, +1/3 for radiation, and −1 for a pure cosmological constant (though the actual value of w is not restricted to just these values).

The usefulness of expressing the RW metric in terms of an observer-dependent coordinate R = a(t) r, which explicitly reveals the dependence of the observed intervals of distance, dR, and time on the curvature induced by the mass-energy content between the observer and R; in the metric, this effect is represented by the proximity of the physical radius R to the cosmic horizon R_h, defined by the relation,

$$\frac{2GM(R_h)}{c^2} = R_h.$$

In this expression, $M(R_h)$ is the mass enclosed within R_h. In terms of ρ, we may also write $R_h = \left(3c^4 / 8\pi G\rho\right)^{1/2}$ or, more simply, $R_h = c / H(t)$ in a flat universe. This is the radius at which a sphere encloses sufficient mass-energy to create a significant time dilation for an observer at the surface relative to the origin of the coordinates.

When the RW metric is written in terms of R, the presence of R_h alters the intervals of time we measure progressively more and more as $R \to R_h$. And since the gravitational time dilation becomes infinite near R_h, it is physically impossible for us to see any process occurring beyond this distance. What this means, of course, is that light emitted beyond Rh is infinitely redshifted and therefore carries no signal.

The motivation for the introduction of these new coordinates was the set of recent observations pointing to a cosmic-horizon radius $R_h(t_0) \approx ct_0$, where to is the inferred current age of the universe. This inference is based on precision measurements of the CMB radiation, which indicate that the universe is extremely flat (i.e., k = 0). Thus ρ is at (or very near) the "critical" value $\rho_0 \equiv 3c^2 H_0^2 8\pi G$.The Hubble Space Telescope Key Project4 on the extragalactic distance scale has measured H with unprecedented accuracy, finding a current value $H_0 \equiv H(t_0) = 71 \pm 6\,\text{km}\,s^{-1}\,\text{Mpc}^{-1}$. It is straightforward to show that with this $H_0 \equiv R_0 \approx ct_0$ (specifically, 13.4 versus 13.7 billion light years). But this empirical result is very peculiar because the FRW equations predict that $\dot{R}_h = c$ only for the very special equation of state $w = -1/3$ which, however, does not appear to be consistent with any of the known constituents of the universe, including a cosmological constant with $w = w_\Lambda \equiv -1$.

As a brief aside, it is interesting to note that de Sitter's own metric (for a universe containing only a cosmological constant) was first written in terms of observer dependent coordinates, though this is no longer widely known, and everyone now uses the form of the RW metric written in terms of r and t only. Of course, de Sitter's work was completed prior to the introduction of the co-moving coordinates (ct, r, θ, ϕ) several years later. A physical motivation for how one might arrive at de Sitter's metric from an application of Schwarzschild's solution to a uniform infinite medium may be understood in terms of Birkhoff's theorem and its corollary.

Given the significance of the fact that $R_h(t_0) \approx ct_0$, the condition $\dot{R}_h = c$ is then always true, guaranteeing the equality (or near equality) of R_h and ct. However, if the equation of state is time dependent, ω need not be exactly −1/3 in order to achieve the result $R_h(t_0) = ct_0$ in the current universe. Indeed, Type Ia supernova data seem to suggest a value of w smaller than this, which motivates us to consider a broader range of constituents in the Universe in order to fully interrogate the available data.

General Coordinate Transformation

In order to write the RW metric in terms of observer-dependent coordinates, we begin with the physical radius,

$$R \equiv a(t) r.$$

However, it will become evident later that to simplify the mathematical transformations, it is convenient to recast Equation $ds^2 = c^2 dt^2 - a^2(t)\left[dr^2\left(1 - kr^2\right)^{-1} + r^2\left(d\theta^2 \sin^2\theta d\phi^2\right)\right]$. using a new function f(t), where,

$$a(t) = e^{f(t)}.$$

In that case,

$$ds^2 = c^2dt^2 - e^{2f(t)}\left[dr^2 + r^2 d\Omega^2\right],$$

where,

$$d\Omega^2 \equiv d\theta^2 + \sin^2\theta d\phi^2.$$

Now,

$$r = \mathrm{Re}^{-f},$$

and therefore,

$$dr = e^{-f}\left[dR - R\dot{f}\,dt\right].$$

So that,

$$dr^2 = e^{-2f}\left[dR^2 + \left(\frac{R\dot{f}}{c}\right)^2 c^2dt^2 - 2\left(\frac{R\dot{f}}{c}\right)c\,dt\,dR\right].$$

Collecting terms, we can now write the metric as,

$$ds^2 = \left[1 - \left(\frac{R\dot{f}}{c}\right)^2\right]c^2dt^2 + 2\left(\frac{R\dot{f}}{c}\right)c\,dt\,dR - dR^2 - R^2 d\Omega^2,$$

where upon, completing the square, we see that,

$$ds^2 = \Phi\left[c\,dt + \left(\frac{R\dot{f}}{c}\right)\Phi^{-1}dR\right]^2 - \Phi^{-1}dR^2 - R^2 d\Omega^2.$$

For convenience we have also defined the quantity,

$$\Phi \equiv 1 - \left(\frac{R\dot{f}}{c}\right)^2,$$

which appears frequently in the metric coefficients. We can easily see that the radius of the cosmic horizon for the observer at the origin is,

$$R_h \equiv c / \dot{f}$$

so that, in effect, the function,

$$\Phi \equiv 1 - \left(\frac{R}{R_h}\right)^2$$

Signals the dependence of the metric on the proximity of the observation radius R to the maximum observable distance R_h. Equation $ds^2 = \Phi\left[c\,dt + \left(\frac{R\dot{f}}{c}\right)\Phi^{-1}dR\right]^2 - \Phi^{-1}dR^2 - R^2 d\Omega^2$. thus becomes,

$$ds^2 = \Phi\left[c\,dt + \left(\frac{R}{R_h}\right)\Phi^{-1}dR\right]^2 - \Phi^{-1}dR^2 - R^2 d\Omega^2.$$

This metric is already in a form we can use to examine the behavior of the cosmological spacetime at any radius R. However, sometimes (e.g., as in the de Sitter metric) it may be useful to "complete" the transformation by introducing a new time coordinate T, such that,

$$\Phi^{-1/2}\frac{cdT}{\eta(t,R)} \equiv \Phi^{1/2}c\,dt + \left(\frac{R}{R_h}\right)\Phi^{-1/2}dR,$$

where $\eta(t,R)$ is an integrating factor selected to guarantee that dT is an exact differential. There are possibly an infinite number of solutions for $\eta(t,R)$, but only one is dimensionless (ensuring that T has dimensions of time), while completely diagonalizing the metric. With this substitution,

$$ds^2 = \Phi^{-1}\frac{c^2 dT^2}{\eta^2(t,R)} - \Phi^{-1}dR^2 - R^2 d\Omega^2.$$

But in order for dT to be an exact differential, T must satisfy the following condition:

$$\frac{\partial^2 T}{\partial R\,\partial t} = \frac{\partial^2 T}{\partial t\,\partial R},$$

which means that $\eta(t,R)$ must be a solution to the equation,

$$\frac{\partial}{\partial R}\left[\Phi\,\eta(t,R)\right] = \frac{\partial}{\partial t}\left[\left(\frac{R}{cR_h}\right)\eta(t,R)\right].$$

However, the coordinate T has only limited applicability because it is rarely possible to integrate dT starting from the big bang at t = 0 to the present.

Let us now examine the behavior of the interval ds connecting any arbitrary pair of spacetime events at R. For an interval produced at R by the advancement of time only (with dR = dΩ = 0), Equation $ds^2 = \Phi\left[c\,dt + \left(\frac{R}{R_h}\right)\Phi^{-1}dR\right]^2 - \Phi^{-1}dR^2 - R^2 d\Omega^2$ gives,

$$ds^2 = \Phi c^2 dt^2.$$

Thus, if we were to make a measurement a fixed distance R away from us, the time interval dt corresponding to any measurable (non-zero) value of ds must go to infinity as $R \to R_h$. (In the context of black-hole physics, we recognize this effect as the divergent gravitational redshift measured by a static observer outside the event horizon.) This result is generic to all cosmologies though, as we shall see, in some cases R_h lies beyond the distance ct_0 light has traveled since the big bang and is therefore not an observable quantity. What does change from case to case, however, are the constituents of the Universe, which directly determine w, and therefore the function f and the horizon's radius R_h.

But we can already see directly from the FRW equations how Rh should behave for any given value of w. It is straightforward to demonstrate from Equations $H^2 \equiv \left(\frac{\dot{a}}{a}\right)^2 = \frac{8\pi G}{3c^2}\rho - \frac{kc^2}{a^2}$, – $\dot{\rho} = -3H(\rho + p)$, that,

$$\dot{R}_h = \frac{3}{2}(1+w)c.$$

Thus, R_h is an increasing function of cosmic time t for any cosmology with w > −1. It is fixed only for de Sitter, in which ρ is a cosmological constant and w = −1. In addition, there is clearly a demarcation at w = −1/3.When w < −1/3, R_h increases more slowly than light speed, and therefore our universe would be delimited by this horizon because light would have traveled a distance ct_0 greater than $R_h(t_0)$ since the big bang. On the other hand, R_h is always greater than ct when w > −1/3, and our observational limit would then simply be set by the light travel distance ct_0.

Specific Cosmologies

The De Sitter Universe (W= −1)

De Sitter himself published the earliest form of his metric using the coordinate R, though he was apparently not aware of its full significance. But as a simple, initial application of the method let us repeat this calculation by considering a universe containing only a cosmological constant. In this case,

$$a(t) = e^{H0t},$$

for which,

$$ds^2 = c^2dt^2 - e^{2H0t}\left[dr^2 + r^2d\Omega^2\right].$$

Clearly,

$$f(t) = \text{In}\left[a(t)\right] = H_0t,$$

So,

$$\dot{f} = H_0.$$

Thus, inserting this form of $\dot{f}$ into Equation $R_h \equiv c/\dot{f}$, we obtain,

$$R_h = c/H_0.$$

As we saw in Equation $\dot{R}_h = \frac{3}{2}(1+w)c.$, the de Sitter metric is unique among the various cosmologies in that the density ρ is constant, and therefore the radius R_h is fixed for all cosmic time t.

However, for all the other cosmologies we will consider below, $w > -1$, so $\dot{R}_h > 0$. This means, of course, that whereas an observer can choose radii R that are always smaller than R_h in de Sitter, the same is not true when $\dot{R}_h > 0$, since for any given R, $R_h > R$ when $t \to 0$.

An important consequence of this distinction is that the interval ds is then imaginary at early times (since $R_h < R$ for all but the de Sitter metric, and dT can therefore not be integrated starting at t = 0. But for de Sitter, we see from Equation $ds^2 = \Phi\left[c\,dt + \left(\frac{R}{R_h}\right)\Phi^{-1}dR\right]^2 - \Phi^{-1}dR^2 - R^2 d\Omega^2$. (since R_h is independent of time) that,

$$\Phi^{1/2} c\,dT \equiv \Phi^{1/2}\left[c\,dt + \left(\frac{R}{R_h}\right)\Phi^{-1}dR\right].$$

A comparison with Equation $\Phi^{-1/2}\frac{cdT}{\eta(t,R)} \equiv \Phi^{1/2} c\,dt + \left(\frac{R}{R_h}\right)\Phi^{-1/2}dR$, tells immediately that the integrating factor $\eta(t,R)$ in the case of de Sitter is simply equal to Φ^{-1}, and a straightforward integration of Equation $\Phi^{1/2} c\,dT \equiv \Phi^{1/2}\left[c\,dt + \left(\frac{R}{R_h}\right)\Phi^{-1}dR\right]$. thus gives,

$$T(t,R) = t - \frac{1}{2H_0}\operatorname{In}\Phi.$$

At the origin, T is of course equal to t, but since the gravitationally-induced dilation increases with R, T also includes the additional redshift seen at progressively greater distances from the observer. Note, however, that for dR = dΩ = 0, dT = dt, so both of these time intervals diverge for a finite ds as $R \to R_h$.

Written in terms of T and R, the de Sitter metric becomes,

$$ds^2 = \Phi c^2 dT^2 - \Phi^{-1}dR^2 - R^2 d\Omega^2,$$

Which is the form originally presented by de Sitter.

A Cosmology with R_h (t) = ct (i.e., w = −1/3)

This is the case in which $\dot{R}_h = c$ in Equation $\dot{R}_h = \frac{3}{2}(1+w)c.$. Aside from being one of the simplest models of the universe that one can construct from the FRW equations, it is also motivated by the observed fact that $R_h(t_0) \approx ct_0$ at the present time. If w is always −1/3, then it would not be surprising to see $R_h(t_0) = ct_0$ now, since this condition would have been true from the beginning.

But though this may seem like a highly idealized model that bears minimal relevance to the real universe, it is actually singularly significant because an equation of state w = −1/3 is the only one for which the current age, to, of the universe can equal the light-crossing time, $t_h \equiv$ Rh/c. For any other cosmology with w < −1/3, t_0 must be greater than R_h/c. We shall return to this shortly. For now, let us consider the transformation of coordinates when w = −1/3, for which,

$$a(t) = H_0 t.$$

Then,

$$f(t) = \ln[a(t)] = \ln(H_0 t),$$

and,

$$\dot{f} = \frac{1}{t}.$$

We therefore confirm that $R_h = ct$ in this case, and from Equation $ds^2 = \Phi\left[c\,dt + \left(\frac{R}{R_h}\right)\Phi^{-1}dR\right]^2 - \Phi^{-1}dR^2 - R^2 d\Omega^2$, we obtain the metric,

$$ds^2 = \Phi\left[c\,dt + \left(\frac{R}{ct}\right)\Phi^{-1}dR\right]^2 - \Phi^{-1}dR^2 - R^2 d\Omega^2,$$

with,

$$\Phi = 1 - \left(\frac{R}{ct}\right)^2.$$

As was the case with de Sitter, the cosmic time dt diverges for a measurable line element as $R \to R_h$ (which, however, is equal to ct here).

We may also write the metric in terms of R and T. A dimensionless solution to Equation $\frac{\partial}{\partial R}[\Phi\,\eta(t,R)] = \frac{\partial}{\partial t}\left[\left(\frac{R}{cR_h}\right)\eta(t,R)\right]$ is:

$$\eta(t,R) = \exp\left\{\frac{1}{2}\left(\frac{R}{ct}\right)^2\right\},$$

which leads to the metric,

$$ds^2 = e^{-(R/R_h)^2}\Phi^{-1}c^2 dT^2 - \Phi^{-1}dR^2 - R^2 d\Omega^2.$$

This form of the metric has only a limited applicability, given that $ds^2 < 0$ when $R_h < R$ and $dR = d\Omega = 0$.

Radiation Dominated Universe (w = +1/3)

The real Universe may indeed contain a cosmological constant, in which case it approaches a de Sitter space time asymptotically, given that its other constituents have a density ρ that drops off as a(t) increases. But let us now move beyond the simple de Sitter application, and the special case with w = −1/3, and consider other cosmological phases that may have emerged since the big bang.

In the very beginning, when radiation dominated the equation of state (with $w = w_{rad} \equiv +1/2$), the expansion parameter was given as,

$$a(t) = (2H_0 t)^{1/2}.$$

In that case,

$$f(t) = \text{In}[a(t)] = \frac{1}{2}\text{In}(2H_0 t),$$

and,

$$\dot{f} = \frac{1}{2t}.$$

Therefore R_h = 2ct. In this case, the visible Universe (extending out to ct) never reaches R_h, and the metric is:

$$ds^2 = \Phi\left[c\,dt + \left(\frac{R}{2ct}\right)\Phi^{-1}dR\right]^2 - \Phi^{-1}dR^2 - R^2\,d\Omega^2,$$

with,

$$\Phi = 1 - \left(\frac{R}{2ct}\right)^2.$$

Thus, measurements made at a fixed R and Ω still produce a gravitationally-induced dilation of dt as R increases, but this effect never becomes divergent within that portion of the Universe (i.e., within ct_0) that remains observable since the big bang.

For completeness, we note that a dimensionless solution for $\eta(t,R)$ from Equation,

$$\frac{\partial}{\partial R}[\Phi\,\eta(t,R)] = \frac{\partial}{\partial t}\left[\left(\frac{R}{cR_h}\right)\eta(t,R)\right]. \text{ is:}$$

$$\eta(t,R) = 1,$$

and the metric written in terms of R and T is therefore,

$$ds^2 = \Phi^{-1}c^2dT^2 - \Phi^{-1}dR^2 - R^2d\Omega^2.$$

The same restrictions as before pertain to T since here also $R_h < R$ at early times, which would make ds imaginary for $dR = d\Omega = 0$ back then.

Matter Dominated Universe (w = 0)

When ρ is dominated by matter with $w \approx w_{\text{matter}} \equiv 0$, the expansion factor grows according to,

$$a(t) = \left(\frac{3}{2}H_0 t\right)^{2/3}.$$

Therefore,

$$f(t) = \text{In}\left[a(t)\right] = \frac{2}{3}\text{In}\left(\frac{2}{3}H_0 t\right),$$

and,

$$\dot{f} = \frac{2}{3t}.$$

In this case, the radius of the cosmic horizon is R_h = (3/2) ct and the metric becomes,

$$ds^2 = \Phi\left[c\,dt + \left(\frac{R}{3ct/2}\right)\Phi^{-1}dR\right]^2 - \Phi^{-1}dR^2 - R^2 d\Omega^2,$$

With,

$$\Phi = 1 - \left(\frac{R}{3ct/2}\right)^2.$$

The situation is similar to that for a radiation dominated universe, in that R_h always recedes from us faster than light speed. Although dilation is evident with increasing R, curvature alone does not produce a divergent redshift.

A dimensionless integrating factor $\eta(t,R)$ is,

$$\eta(t,R) = \sqrt{1 + \frac{1}{2}\left(\frac{R}{R_h}\right)^2}.$$

Thus, in terms of R and T,

$$ds^2 = \Phi^{-1}\left[1 + \frac{1}{2}\left(\frac{R}{R_h}\right)^2\right]^{-1} c^2 dT^2 - \Phi^{-1}dR^2 - R^2 d\Omega^2.$$

Universe Delimited by the Cosmic Horizon (−1 < w < −1/3)

In general, R_h = (3/2) (1 + w) ct for a universe with constant w. Except for the special cases w = −1 and w = −1/3, the metric may therefore be written as,

$$ds^2 = \Phi\left[c\,dt + \left(\frac{2R}{3(1+w)ct}\right)\Phi^{-1}dR\right]^2 - \Phi^{-1}dR^2 - R^2 d\Omega^2,$$

With,

$$\Phi = 1 - \left(\frac{2R}{3(1+w)ct}\right)^2.$$

A cosmology with w < −1/3 is therefore delimited by the horizon in the sense that our measurements—extending over a distance cto—would have probed regions of space time that include $R_h(t_0)$ today.

For example, in what may be a close approximation to the Universe we have currently, the line element for w = −2/3 is,

$$ds^2 = \Phi\left[c\,dt + \left(\frac{2R}{ct}\right)\Phi^{-1}dR\right]^2 - \Phi^{-1}dR^2 - R^2 d\Omega^2,$$

Where $\Phi = 1-(2R/ct)^2$. In this Universe, the cosmic horizon would lie at a radius from us only half of the total distance traversed by light since the big bang, and we would therefore be sampling phenomena near the gravitational limit.

It is not difficult to see that the general form of the metric written in terms of R and T is:

$$ds^2 = \Phi^{-1}\left[1+\frac{1}{2}(1+3w)\left(\frac{R}{R_h}\right)^2\right]^{(3w-1)/(3w+1)} c^2 dT^2 - \Phi^{-1}dR^2 - R^2 d\Omega^2.$$

Structure Formation

The growth of seed fluctuations into cosmic structure can be summarised into three main phases:

- Between inflation and the release of the cosmic microwave background.
- Between the release of the cosmic microwave background and the formation of the first stars and galaxies.
- After the formation of the first stars and galaxies.

Between inflation and the release of the cosmic microwave background (t <1 sec to t =380,000 years): After the end of inflation, the Universe consisted of a more or less uniform bath of fundamental particles, like quarks, electrons and their anti-particles. There were also neutrinos, photons (the particles of light) and dark matter particles, an unknown type of massive particle that does not interact with photons and is therefore dark (as it does not emit light). At this time there was slightly more matter than anti-matter, but as the particles collided with their anti-particles they annihilated, leaving the Universe dominated by particles, and anti-matter disappeared. Quarks then teamed up in trios, forming protons or neutrons – the constituents of atomic nuclei as we know them today. This all happened within the first second after the Big Bang. About three minutes after the Big Bang, protons and neutrons had combined to form the nuclei of hydrogen and helium.

The density and temperature of particles in the early Universe were extremely high, and collisions between the particles were very frequent. Cosmologists refer to this by saying that ordinary matter (such as electrons, protons, neutrons and the few atomic nuclei that had formed by then) was tightly coupled to the photons. Because of these frequent interactions, photons could not travel freely: the Universe was opaque. Besides, ordinary matter is subject to gravity and ideally any denser region – such as the seed fluctuations that were present at the end of inflation – would draw more matter from their surroundings, growing denser and more massive. However, ordinary matter at this epoch was coupled to the photons, and the radiation pressure of photons pushes away any concentration of matter that may be created under the effect of gravity. This phenomenon prevents any fluctuations in the distribution of ordinary matter to grow denser as long as matter is coupled to the photons.

At the same time, dark matter particles were not bound to the photons, since the two species do not interact with one another. This type of dark matter particle is also referred to as cold dark matter because the velocity of these particles is much lower than the speed of light. Hence, fluctuations in the distribution of cold dark matter can grow denser and more massive even before the release of the cosmic microwave background.

Astronomers also refer to hot dark matter, or neutrinos – particles with a very small mass and no electric charge that travel nearly at the speed of light. In the first second of the Universe, neutrinos were coupled to the photons, but these two types of particles decoupled immediately after. Since they do not interact with light during most of the Universe's history, neutrinos can be considered as a type of dark matter, and since their velocity is close to the speed of light, they are regarded as hot dark matter. Fluctuations in the distribution of hot dark matter can grow denser and more massive, but due to their high velocity, these particles tend to dissipate and their fluctuations are damped on small scales so, effectively, only fluctuations on intermediate and large scales can grow.

The growth of primordial fluctuations in hot and cold dark matter gave rise to two completely different distributions of cosmic structure. In hot dark matter models, the first structures to form are the most massive, that subsequently fragment into smaller and smaller structures. This has been discarded on the basis of observations of galaxies in the early Universe: since the first objects that are seen to emerge in cosmic history have low mass, and they gradually evolve into more massive structures, cosmologists have established that the bulk of dark matter in the Universe is cold. However, a small fraction of hot dark matter is present in the Universe as neutrinos. Depending on the mass of neutrinos (which has not been determined yet) the effect of hot dark matter can be more or less evident in the distribution of cosmic structure on different scales, since neutrinos tend to smooth out the formation of small-scale structures.

Between the release of the cosmic microwave background and the formation of the first stars and galaxies (t = 380,000 years to t = a few hundred million years): About 380,000 years after the Big Bang, the Universe had expanded enough so that its density was much lower than at earlier epochs. Likewise, the temperature of the Universe had cooled down from the billions of Kelvin of the first few minutes and had reached about 3000 Kelvin. Protons and electrons could finally combine to form atoms of neutral hydrogen. Electrons disappeared from the view of photons and these two species decoupled from one another. This marked the beginning of the period known as the Dark Ages – a name arising from the fact that there were no individual sources of light, like stars, only clouds of neutral hydrogen.

The decoupling had two effects: Photons were free to propagate across the Universe, which was now largely transparent, and which we observe as the cosmic microwave background (CMB); on the other hand, ordinary matter particles were free to assemble under the effect of gravity. From this moment on, ordinary and dark matter could both react to gravity: denser concentrations of matter (both ordinary and dark) grew denser and more massive. Since dark matter particles had already created a network of dense and empty structure, ordinary matter particles could feel the gravitational attraction from the densest concentrations of dark matter and fall toward them. But ordinary matter could also get rid of energy quite effectively by heating up and emitting radiation, which caused it to sink even further into the already existing regions of high matter density. These processes gave rise to a highly sub-structured network of sheets and filaments of ordinary and dark matter known as the cosmic web, which constitutes the skeleton supporting the later emergence of stars and galaxies. Eventually the densest concentrations gave rise to the first stars, leading to the end of the Dark Ages.

After the formation of the first stars and galaxies (t = a few hundred million years to t = now): A few hundred million years after the Big Bang, the distribution of matter in the Universe had produced very dense knots at the intersections of the sheets and filaments that make up the cosmic web. In these knots, the density of ordinary matter was so high that the formation of stars and galaxies became possible. Eventually the first stars and galaxies sparked into existence and light could escape from them, revealing the distant Universe to telescopes today.

The first stars were formed almost exclusively out of hydrogen and helium and are believed to have been extremely massive (about 100 times the mass of the Sun or more) and to have lived very short lives, exploding soon after their formation as supernovae and releasing their material in the surroundings, triggering the birth of new stellar generations. Later generations included other elements formed in the nuclear furnace of previous stars, and their masses were typically smaller. The first generation of stars formed in relatively low-mass galaxies. Massive galaxies, and even more massive structures such as galaxy clusters, formed later.

How did the Formation of Structure Affect the Cosmic Microwave Background?

The birth of the first stars and galaxies had an interesting effect on the cosmic microwave background (CMB) photons. Ultraviolet radiation released by these objects ionised hydrogen atoms, turning them back into protons and electrons. This created a series of expanding bubbles of ionised gas – a bit like the holes in Swiss cheese – and within a few hundred million years these bubbles had merged and the entire Universe was ionised again, a period of time termed re-ionisation.

The CMB photons were affected by the re-ionisation; they were scattered off the free electrons in the re-ionised Universe, washing out some of the primordial fluctuations in the CMB as we observe it today. Since this happened when the Universe was already mature and had reached a substantial size, the effect of re-ionisation can be detected in the fluctuations of the CMB on large scales. This effect is expressed in terms of the 'opacity', which describes the average density of free electrons that are present along the line of sight between an observer (in this case, the telescope on board Planck) and the CMB. This parameter also provides a tool to estimate when the first stars formed.

How is the History of Cosmic Structure Encoded in the Cosmic Microwave Background and Power Spectrum?

The variations in the density of matter at the time when the cosmic microwave background (CMB) formed derive from the seed fluctuations that were produced at the end of inflation and can be deciphered by looking at the power spectrum for cosmic structure in the Universe at a range of scales.

At scales smaller than about one degree – or twice the size of the full Moon on the sky – the graph shows the imprint and oscillation pattern of sound waves that were present in the fluid of ordinary matter and radiation in the very early Universe, before the CMB was released. At this epoch, ordinary matter was tightly coupled to the photons, and the radiation pressure of photons pushed away any concentration of matter that might have been created under the effect of gravity.

The interplay between gravity, which pulled together the fluid of matter and radiation, and the radiation pressure, which pushed it away, caused a series of rhythmical compressions and rarefactions everywhere in the fluid. This results in the pattern of sound waves that is visible in the central part of the power spectrum graph. Since gravity is caused by both dark and ordinary matter particles, but the radiation pressure of photons is only experienced by ordinary matter (because dark matter particles are not coupled to photons), the shape of these oscillations contains information about the amount of ordinary matter relative to the amount of dark matter. As dark matter was not bound to the photons, any concentration of dark matter could grow denser and denser even before the release of the CMB. The relative contribution of ordinary matter particles (also referred to as baryons) to the overall cosmic budget is expressed in terms of the 'Omega_b' parameter, where b stands for baryons, and the relative contribution of cold dark matter particles is expressed in terms of the 'Omega_c' parameter, where c stands for cold. The 'cold' in cold dark matter refers to the low speed of these particles ('warm' dark matter particles move at higher speed and 'hot' dark matter particles move at the speed of light).

While gravity pulls matter together to form structures, the expansion of the Universe may counteract this effect and hamper the formation of cosmic structure. For this reason, the amount of fluctuations in the Universe depends also on the speed of cosmic expansion, and this quantity can be extracted from the shape of the oscillations in the power spectrum of the CMB. The speed of the Universe is expressed in terms of the Hubble constant, H_o, which quantifies the expansion of the Universe at present time.

What does the Cosmic Microwave Background Tell us about the Overall 'Shape' of the Universe?

The CMB holds clues to the nature and distribution of structure in the Universe, and the average density of this matter plays a key role in determining the geometry of the Universe. The geometry of the Universe can take on one of three shapes: it can be curved like the surface of a ball and finite in extent (positively curved); curved like a saddle and infinite in extent (negatively curved), or it can be flat and infinite. The geometry and density of the Universe are related in such a way that, if the average density of matter in the Universe is found to be less than the so-called critical density (roughly equal to 6 hydrogen atoms per cubic metre) the Universe is open and infinite. If the density is greater than the critical density the Universe is closed and finite. If the density just equals the critical density, the Universe is flat.

Cosmologists study the relative sizes of the oscillations of the fluid of matter and radiation at the time the CMB was released to learn more about the shape of the Universe. The oscillations translate into regions of higher and lower temperature on the CMB map, and contain information about the amount of particles present. More specifically, the shape of the Universe can be determined by looking at where the first of these oscillations appears in the power spectrum.

The location of the first oscillation corresponds to a specific size in the early Universe called the sound horizon – the maximum distance that a sound wave could have crossed from the Big Bang until the time of the CMB release. To cosmologists, the sound horizon works like a standard measure of known length. By measuring its length in the temperature fluctuations of the CMB, it is possible to determine if the Universe is flat or curved. This is expressed in terms of the parameter 'Omega_K' and is equal to zero for exactly flat space.

Cosmological Components

Dark Energy

Dark energy is the repulsive force that is the dominant component (69.4 per cent) of the universe. The remaining portion of the universe consists of ordinary matter and dark matter. Dark energy, in contrast to both forms of matter, is relatively uniform in time and space and is gravitationally repulsive, not attractive, within the volume it occupies. The nature of dark energy is still not well understood.

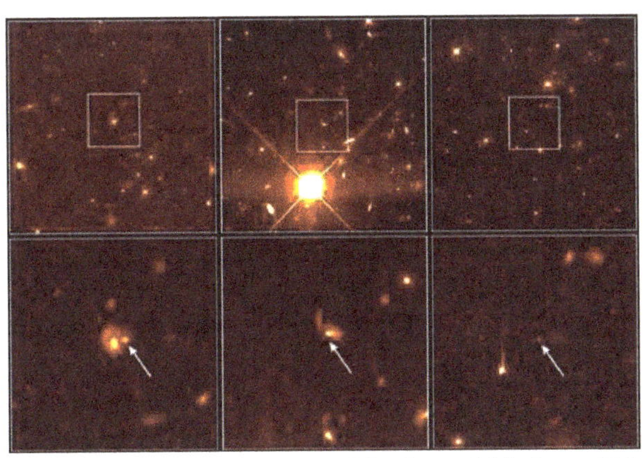

Three distant Type Ia supernovas, as observed by the Hubble Space Telescope in 1997.

Since Type I_a supernovas have the same luminosity, they are used in measuring dark energy and its effects on the expansion of the universe. The bottom images are details of the upper wide views. The supernovas at left and centre occurred about five billion years ago; the right, seven billion years ago.

A kind of cosmic repulsive force was first hypothesized by Albert Einstein in 1917 and was represented by a term, the "cosmological constant," that Einstein reluctantly introduced into his theory of general relativity in order to counteract the attractive force of gravity and account for a universe that was assumed to be static (neither expanding nor contracting). After the discovery in the 1920s by American astronomer Edwin Hubble that the universe is not static but is in fact expanding, Einstein referred to the addition of this constant as his "greatest blunder." However, the measured amount of matter in the mass-energy budget of the universe was improbably low, and thus some unknown "missing component," much like the cosmological constant, was required to make up the deficit. Direct evidence for the existence of this component, which was dubbed dark energy, was first presented in 1998.

Dark energy is detected by its effect on the rate at which the universe expands and its effect on the rate at which large-scale structures such as galaxies and clusters of galaxies form through gravitational instabilities. The measurement of the expansion rate requires the use of telescopes to measure the distance (or light travel time) of objects seen at different size scales (or redshifts) in the history of the universe. These efforts are generally limited by the difficulty in accurately measuring astronomical distances. Since dark energy works against gravity, more dark energy accelerates the universe's expansion and retards the formation of large-scale structure. One technique for measuring the expansion rate is to observe the apparent brightness of objects of known luminosity like Type Ia supernovas. Dark energy was discovered in 1998 with this method by two international teams that included American astronomers Adam Riess and Saul Perlmutter and Australian astronomer Brian Schmidt. The two teams used eight telescopes including those of the Keck Observatory and the MMT Observatory. Type Ia supernovas that exploded when the universe was only two-thirds of its present size were fainter and thus farther away than they would be in a universe without dark energy. This implied the expansion rate of the universe is faster now than it was in the past, a result of the current dominance of dark energy. (Dark energy was negligible in the early universe.)

Studying the effect of dark energy on large-scale structure involves measuring subtle distortions in the shapes of galaxies arising from the bending of space by intervening matter, a phenomenon known as "weak lensing". At some point in the last few billion years, dark energy became dominant in the universe and thus prevented more galaxies and clusters of galaxies from forming. This change in the structure of the universe is revealed by weak lensing. Another measure comes from counting the number of clusters of galaxies in the universe to measure the volume of space and the rate at which that volume is increasing. The goals of most observational studies of dark energy are to measure its equation of state (the ratio of its pressure to its energy density), variations in its properties, and the degree to which dark energy provides a complete description of gravitational physics.

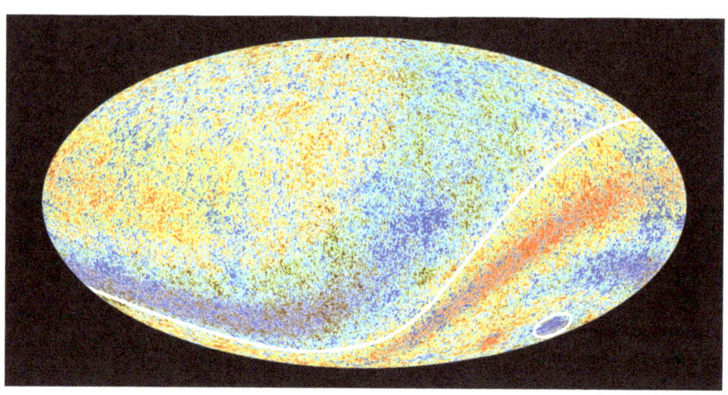

A full-sky map produced by the Wilkinson Microwave Anisotropy Probe (WMAP) showing cosmic background radiation, a very uniform glow of microwaves emitted by the infant universe more than 13 billion years ago. Colour differences indicate tiny fluctuations in the intensity of the radiation, a result of tiny variations in the density of matter in the early universe. According to inflation theory, these irregularities were the "seeds" that became the galaxies. WMAP's data support the big bang and inflation models.

In cosmological theory, dark energy is a general class of components in the stress-energy tensor of the field equations in Einstein's theory of general relativity. In this theory, there is a direct correspondence between the matter-energy of the universe (expressed in the tensor) and the shape of space-time. Both the matter (or energy) density (a positive quantity) and the internal pressure contribute to a component's gravitational field. While familiar components of the stress-energy tensor such as matter and radiation provide attractive gravity by bending space-time, dark energy causes repulsive gravity through negative internal pressure. If the ratio of the pressure to the energy density is less than $-1/3$, a possibility for a component with negative pressure, that component will be gravitationally self-repulsive. If such a component dominates the universe, it will accelerate the universe's expansion.

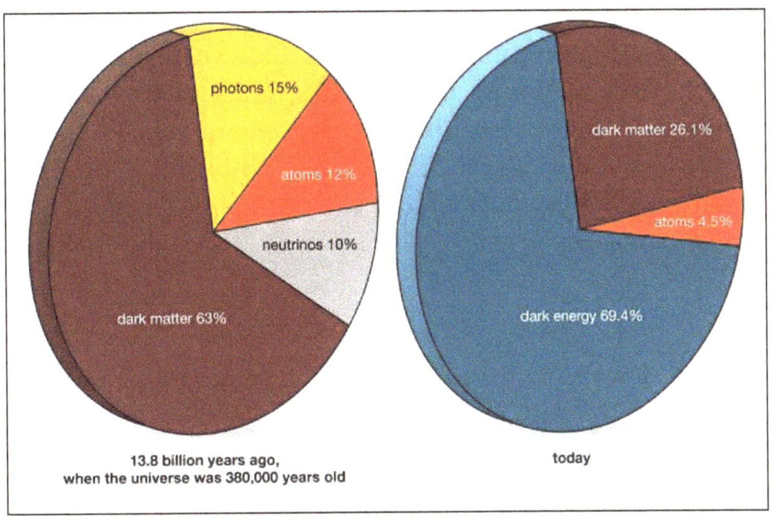

Matter-energy content of the universe.

The simplest and oldest explanation for dark energy is that it is an energy density inherent to empty space, or a "vacuum energy." Mathematically, vacuum energy is equivalent to Einstein's cosmological constant. Despite the rejection of the cosmological constant by Einstein and others, the modern understanding of the vacuum, based on quantum field theory, is that vacuum energy arises naturally from the totality of quantum fluctuations (i.e., virtual particle-antiparticle pairs that come into existence and then annihilate each other shortly thereafter) in empty space. However, the observed density of the cosmological vacuum energy density is $\sim 10^{-10}$ ergs per cubic centimetre; the value predicted from quantum field theory is $\sim 10^{110}$ ergs per cubic centimetre. This discrepancy of 10^{120} was known even before the discovery of the far weaker dark energy. While a fundamental solution to this problem has not yet been found, probabilistic solutions have been posited, motivated by string theory and the possible existence of a large number of disconnected universes. In this paradigm the unexpectedly low value of the constant is understood as a result of an even greater number of opportunities (i.e., universes) for the occurrence of different values of the constant and the random selection of a value small enough to allow for the formation of galaxies (and thus stars and life).

Another popular theory for dark energy is that it is a transient vacuum energy resulting from the potential energy of a dynamical field. Known as "quintessence," this form of dark energy would vary in space and time, thus providing a possible way to distinguish it from a cosmological constant. It is also similar in mechanism (though vastly different in scale) to the scalar field energy invoked in the inflationary theory of the big bang.

Another possible explanation for dark energy is topological defects in the fabric of the universe. In the case of intrinsic defects in space-time (e.g., cosmic strings or walls), the production of new defects as the universe expands is mathematically similar to a cosmological constant, although the value of the equation of state for the defects depends on whether the defects are strings (one-dimensional) or walls (two-dimensional).

There have also been attempts to modify gravity to explain both cosmological and local observations without the need for dark energy. These attempts invoke departures from general relativity on scales of the entire observable universe.

A major challenge to understanding accelerated expansion with or without dark energy is to explain the relatively recent occurrence (in the past few billion years) of near-equality between the density of dark energy and dark matter even though they must have evolved differently. (For cosmic structures to have formed in the early universe, dark energy must have been an insignificant component.) This problem is known as the "coincidence problem" or the "fine-tuning problem." Understanding the nature of dark energy and its many related problems is one of the most formidable challenges in modern physics.

Dark Fluid

The dark fluid framework generically branches into a continuous spectrum of theories with DE and DM effects, including the f(R) gravity, tensor-vector-scalar-like theories, Einstein-Aether, and vΛ theories as limiting cases. When the vector field degenerates into a pure scalar field, we obtain the physics for quintessence.

Background Cosmology

Consider background cosmology in the FRW flat metric,

$$ds^2 = dt^2 - a^2(t)\left(dx^2 + dy^2 + dz^2\right).$$

First, the scalar field follows an EOM exactly as a quintessence,

$$\ddot{\varphi} + 3H\dot{\varphi} = -\left(\Lambda_0 F' + \frac{3\alpha'}{2}H^2\right)\left(2N^2\varphi\right),$$

so φ tracks the Hubble rate $H = \dot{a}/a$. Here we introduce an auxiliary function b(λ) or α(λ) defined by $b(\lambda) - 1 \equiv \frac{\alpha(\lambda)}{2} \equiv \frac{c_1 + c_3 + 3c_2}{2} = -\lambda^{-1}$ for our choice of coefficients $c_2 = -\frac{1}{3\lambda}$, where $\lambda = \varphi^2$. The vector field EOM gives the Lagrange multiplier $L^* = \partial_t(\alpha H) + \frac{1 + c_\varphi^2}{N^2}\dot{\varphi}^2$. Since L^* and $N^2\Lambda_0 F'$ can be thought as the masses of the scalar and vector fields, hence our model describes effectively an unstable slowly decaying dark particle. To be more explicit, consider the approximation $\alpha' \to 0$ and $\Lambda_0 F \sim 9\varphi^2$ for large φ in our model with k = 3. The scalar EOM becomes a damped harmonic equation,

$$\ddot{\varphi} + 3H\dot{\varphi} = -18N^2\Lambda_0\varphi,$$

which gives the approximate solution,

$$9\Lambda_0\varphi^2 \approx 3\Omega_0 H^2 \cos^2\left(Nt\sqrt{18\Lambda_0}\right),$$

$$\frac{1}{2N^2}\dot{\varphi}^2 \approx 3\Omega_0 H^2 \sin^2\left(Nt\sqrt{18\Lambda_0}\right),$$

which is rigorous in the matter-dominated era when $H^{-1} = \frac{3}{2}t$ where $\Omega_0 \sim O(1)$ is a constant, determined by the initial condition of phiv. Taking the average over the rapid oscillations, we have:

$$(F-9)\Lambda_0 \approx \frac{1}{2N^2}\dot{\varphi}^2 \approx \frac{3\Omega_0}{2}H^2$$

at high redshift. A numerical example of the oscillating scalar field is given in figure.

The modified 00 term of the Einstein equation becomes,

$$3H^2 = 8\pi G\left[\bar{\rho} + \bar{\rho}_{DS}\right],$$

$$\bar{\rho}_{DS} \equiv \frac{1}{8\pi G\left(1+\frac{\alpha}{2}\right)}\left[\frac{1}{2N^2}\dot{\varphi}^2 + \Lambda_0 F - \alpha 4\pi G\bar{\rho}\right]$$

Where $\bar{\rho}_{DS}$ and $\bar{\rho}$ are the (background) energy densities of the dark sector (DM plus DE) and the baryon-radiation fluid, respectively. At high redshift when $F \to \lambda$ we get a DM-like effect, and $\left[\frac{1}{2N^2}\dot{\varphi}^2 + \Lambda_0\varphi^2\right]$ behaves like a CDM of the Broglie frequency $\sqrt{2N^2\Lambda_0}$. The DM-like effect is sub-dominant at very high redshift, e.g., radiation era or Big Bang nucleosynthesis (BBN), where $\lambda \to \infty$, hence $\alpha \to 0$. Therefore, the BBN constraint is automatically satisfied because the Hubble expansion rate at BBN is equal to that of a radiation-only universe. We also expect the early structure formation to resemble CDM models.

Equivalently, the Einstein equation can be written as:

$$-\left(2\frac{\ddot{a}}{a} + H^2\right) = 8\pi G\left(\bar{p} + \bar{p}_{DS}\right)$$

$$\bar{p}_{DS} \equiv \frac{1}{8\pi G\left(1+\frac{\alpha}{2}\right)}\left[\frac{1}{2N^2}\dot{\varphi}^2 - \Lambda_0 F + 2\alpha' H\varphi\dot{\varphi} - \alpha 4\pi G\bar{p}\right],$$

where the pressure of the baryon-radiation fluid $\bar{p}=0$ in the matter-dominated era. Applying the harmonic approximation $\frac{1}{2N^2}\dot{\varphi}^2 \sim \Lambda_0\left[F(1)\right]$ we find the effective pressure of the dark sector,

$$-\bar{p}_{DS} \sim \frac{9\Lambda_0}{8\pi G\left(1+\frac{\alpha}{2}\right)}.$$

This behaves like a DE with a characteristic scale Λ_0, which is of the same order of magnitude (smaller by a factor of a few) as the observed cosmological constant ~ $(8 \times 10^{-10}\,\mathrm{m\,s^{-2}})^2$. Note that in writing the above equations we have implicitly assumed that the effective DM and effective DE components couple to each other. This can been seen by checking that neither DM nor DE satisfies the conservation law $\bar{\rho}+3H\left(\bar{\rho}+\bar{\rho}\right)=0$, but their sum ρ_{DS} does. Figure show two crude estimators for the equation of state parameter and possible problematic regions.

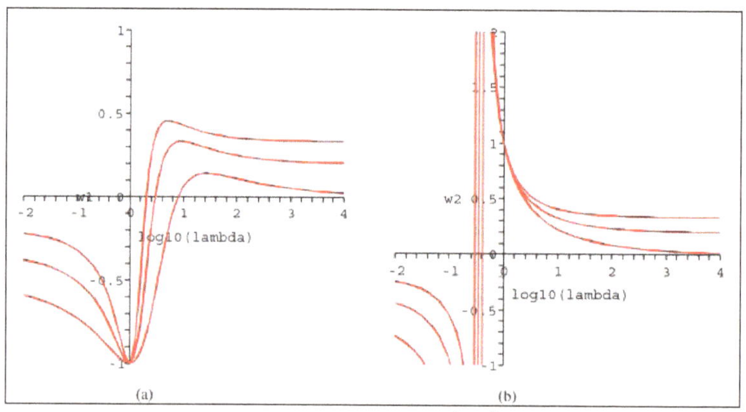

Figure (a): equation of state estimator $w_1 = \frac{\varphi^2 F' - F\left(\varphi^2\right)}{\varphi^2 F' + F\left(\varphi^2\right)}$ vs. $\log_{10}(\lambda)$ for k = 3, 3/2, 2 (bottom to top); note that w1 becomes negative for $\lambda < 10$. (b) Shows the sound of speed estimator $w_2 = \frac{\delta\left[\varphi^2 F' - F\left(\varphi^2\right)\right]}{\delta\left[\varphi^2 F' + F\left(\varphi^2\right)\right]}$. Note that k = 3 has very small sound speed a large λ, but there is a singularity near $\lambda \sim 1/3$.

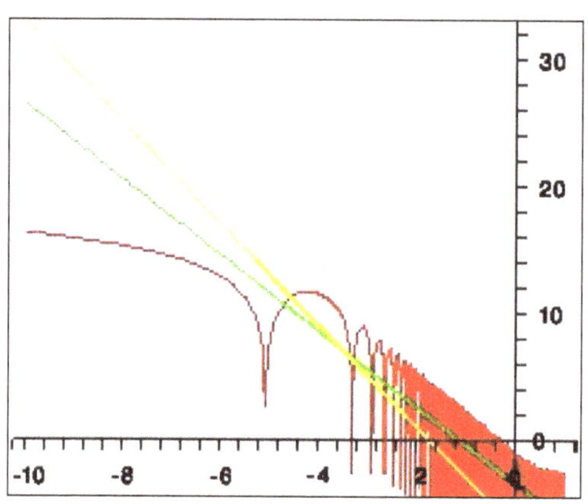

Figure above shows Contributions to the Hubble expansion $\ln\left(H^2 / H_0^2\right)$ from three components as a function of the scale factor in a: baryonic matter (green straight line with an a^{-3} fall-off), radiation (yellow straight line with an a^{-4} fall-off), and dark fluid's potential term $a^2_0 F/3$ (the oscillating curve with a nearly a^{-3} fall-off of its amplitude for the epoch between $1 \gg H_0 t \gg 1/N$). For illustration purposes, we assume parameters $\Omega b = 10\Omega r = 0.03$, and $N = 1000$, an initial condition $\phi = 1.8N$ at $\ln a = -10$. Note the dark fluid is sub-dominate in the radiation era, but dominates the baryonic fluid by a factor of 2 in the matter era, and starts to flat out to a constant amplitude at the present epoch, similar to the cosmological constant effect. More realistic models with a larger N are challenging to integrate numerically (appropriate for studying inflation and DM effects in galaxies), and the results are sensitive to the initial condition of the scalar field. Nevertheless, the dark fluid seems to have the desired properties of in between that of a uniform cosmological constant and that of a CDM.

Static Galaxy Limit

To work out perturbations in static galaxies, remember that in the Newtonian gauge we have only two scalar mode perturbation potentials, Φ and Ψ, which appear in the perturbed metric:

$$ds^2 = (1+2\Phi)dt^2 - a_0^2(1+2\Psi)(dx^2+dy^2+dz^2),$$

Where we will let $a_0 = 1$. As a first study we assume no Hubble expansion.

The vector field EOM in static systems fixes $Æa = g_{ab}Æb = (1+\Phi, 0, 0, 0)$, so the vector field tracks the metric exactly without any freedom in static galaxies.

The 00 component of the Einstein equation becomes a Poisson equation,

$$\sum_{i=x,y,z} -(2\Psi)_{,ii} = 8\pi G(\rho + \rho_{DM} + b\rho_{DE}),$$

$$\rho_{DM} \equiv \frac{\sum_{i=x,y,z}}{8\pi G}[2\lambda\Phi_{,i}]_{,i},$$

Where we use the notation $F_{,i} \equiv \partial iF$, and the dummy index implies covariant or contra-variant derivatives with respect to x, y, z. We use the approximation that the DE part $8\pi Gb\rho_{DE} = \frac{\dot{\varphi}^2}{2N^2} + \Lambda_0 F$ is a negligible source compared to $8\pi G\rho$ from the baryons, and that,

$$-\Psi = \Phi$$

from the spatial cross term of the Einstein equation. The above result is essentially a Poisson equation where the vector field creates an effective DM-like source term ρ_{DM}. Rearrange the terms; the same equation becomes the MOND Poisson equation:

$$\nabla\cdot\left[(1-\lambda)\nabla\Phi\right] = 4\pi G\rho, \quad \lambda \equiv \varphi^2.$$

To see that $1 - \lambda$ can be identified with the MOND μ_M function, first we define a value of the scalar field φ_M such that,

$$F'\big|\lambda = \varphi_M^2 \equiv \frac{|\nabla\Phi|^2}{\Lambda_0}.$$

We find that the scalar field EOM is given as,

$$-c_\varphi^2\nabla^2\varphi = -\left[\Lambda_0 F' - |\nabla\Phi|^2\right]\left(2N^2\varphi\right),$$

where we neglect all time-dependent terms. This equation is similar to the equation of the Yukawa potential with a screening length of c_φ / ω, $(c^2{}_\varphi\nabla^2 - \omega^2)\varphi = 0$. In the simplest case, we adopt $c^2\varphi \to 0$ to kill the Laplacian term $\nabla^2 \equiv \sum_{i=x,y,z} \partial_i \partial_i$, and we find that the equation for the scalar field becomes,

$$k^2\Lambda_0\lambda^{\frac{3}{k}-1}\left(\lambda^{-1/k} - 1\right)^2 \approx |\nabla\Phi|^2$$

for our choice of F(λ), where we neglect the term $\varepsilon^4\varphi^2$ for a lighter expression. For k = 3, the equation can then be easily solved as,

$$\lambda \equiv \varphi^2 \to \varphi_M^2 \approx \left(1 + \frac{x}{3}\right)^{-3}\Bigg|_{x = \frac{|\nabla\Phi|}{\sqrt{\Lambda_0}}}.$$

To see if we recover the properties of MOND function μ_M, we rewrite the solution of the scalar field as,

$$1 - \varphi_M^2 \equiv \mu_M \approx \begin{cases} x, & \text{where } x \equiv \dfrac{|\nabla\Phi|}{\sqrt{\Lambda_0}} \ll 1 \\ 1 - \varepsilon^2 - \left(\dfrac{x}{3}\right)^{-3}, & \text{where } x \gg 1. \end{cases}$$

This is exactly the physics of MOND, if,

$$\sqrt{\Lambda_0} \to a_0$$

is identified with the MOND acceleration scale a_0. In the solar system or strong gravity regime, the modification factor $1 - (x/3)^{-3} \sim 1$ to the Newtonian Poisson equation is small and reduces sharply. In weak gravity, applying the spherical approximation around a dwarf galaxy of mass mb, we have $|\nabla\Phi|^2 / a_0 = Gm_b r^{-2}$, and the rotation curve $V^2{}_{cir}(r) = r\nabla\Phi$. The big success of MOND in dwarf spiral galaxies is to explain their Tully–Fisher relation $V^4{}_{cir}(r)/(Gmb) = a_0 \sim 10^{-10}\, ms^{-2}$ if $\Lambda_0 \sim \left(1 \times 10^{-10}\, ms^{-2}\right)2$, which is of the order of magnitude of the observed amplitude of "the cosmological constant" effect. In the intermediate regime, our μ_M resembles the "standard" $\mu = \frac{x}{\sqrt{1+x^2}}$ function of MOND, so it will fit rotation curves of galaxies very well.

Temporal and Spatial Corrections to Mond: Oscillations and Diffusions

When considering merging systems like galaxy clusters, time-dependent terms $\ddot{\lambda} \sim \dot{\lambda}^2 \sim O(\omega^2)$ re important, where $\omega = O(|k|\sigma)$ is the inverse of the timescale to cross a system of size $|k|^{-1}$ by stars of velocity dispersion σ in unit of the speed of light. There can also be diffusion on small scale due to a pressure-like term $\nabla^2\lambda = -|k|^2\lambda$.

The scalar field EOM becomes:

$$\left[\partial_t^2 - c_\varphi^2\nabla^2 + \left(1-c_\varphi^2\right)\eta\,\partial_t\right]\varphi - \left(\varphi-\varphi_M\right)v^2,$$

Where $v^2 \equiv 2N^2\varphi\Lambda_0 \dfrac{F'-F'_M}{\varphi-\varphi_M} \sim 4N^2\Lambda_0 F''\varphi_M^2$, and $F = F\left(\varphi^2\right), F'_M \equiv F'\left(\varphi^2{}_M\right) = \left(\nabla\Phi\right)^2\Lambda_0$ by definition of φ_M. We also introduce η^{-1} as a damping timescale due to coupling of φ with the Æ vector field. The diffusion term $-c^2\varphi\nabla^2 = c^2\varphi\left|k\right|^2$ is non-zero unless $c^2\varphi = 0$. The scalar field φ then follows the equation of a damped harmonic oscillator with a damping rate $\left(1-c^2\varphi\right)\eta$, a slightly nonlinear restoring force $\sim -v^2\varphi$, and an external force $\sim v^2\varphi_M \sim \left|\nabla\Phi\right|^2\left(2N^2\varphi_M\right)$. Assuming that the correction due to Hubble expansion $O\left(3H^2+2H\eta\right)b'$ is negligible for a very small b', the scalar field φ eventually approaches the MOND-like static solution φ_M. thanks to the damping term with a timescale η^{-1}, which kills any history dependence. Rapid oscillations will likely keep the fluid's time-averaged property close to MOND-like solution as well.

We estimate the oscillation timescale,

$$\sqrt{\frac{\varphi}{\ddot{\varphi}}} \sim v^{-1} = \left(2N\right)^{-1}\left(\Lambda_0 F''\lambda\right)^{-1/2}\Big|_{\lambda=\varphi_M^2} \sim \frac{10^8}{N}\times 300\,\text{yr},$$

which is about 10^9 yr if $N \sim 10$. Here we assume $\varphi_M = \sqrt{\lambda} = O(1) = F''$ for systems of mild gravity ($\sim 10^{-8}\,cm\,s^{-2}$, e.g., clusters; for systems of stronger gravity, the timescale is perhaps longer). In the process of damping, there will be a correction to the MOND μ_M function by the q term, heuristically, $1-\varphi^2 = 1-\varphi^2{}_M$, if,

$$x \to \sqrt{\frac{\left|\nabla\Phi\right|^2}{\Lambda_0} + \frac{q}{2N^2\Lambda_0}},$$

Where $q \equiv \left[\partial^2{}_t - c^2\varphi\nabla^2 + \left(1-c^2\varphi\right)\eta\partial_t\right]$ is an operator. In tidally acting systems, the value for φ will oscillate between its pre-merging value and its equilibrium value.

Models with a small N would not give MOND, e.g., if $N = 1-10, \pi^{\nu-1} \sim (100-10)$ Gyr, then the universe would be too young dynamically to have a precise MOND effect in galaxies because φ would not have enough time to respond to the formation of galaxies. Rather φ would lag behind and might remain close to its cosmological average,

$$\varphi \sim \bar{\varphi},$$

Which would mean a boost of the gravity of the baryon by a constant factor $\left(1-\bar{\varphi}^2\right)^{-1}$ everywhere.

Generic Properties of Dark Fluid

It is still uncertain whether the time-dependent correction and a possible diffusion term are enough to help MOND to explain the Bullet clusters. However, it seems robust that the *dark fluid*—described by the field $Æ^a\varphi(\lambda)$—is generally out of phase from the baryonic fluid. The dynamics of galaxies inside the dark fluid are governed by five variables, two for the scalar field $\varphi = \sqrt{\lambda}$, the metric or potential $\Phi = -\Psi = \dfrac{g_{00}-1}{2} = Æ_0 - 1$ three for the vector field perturbation part $Æ_i = Æ_i - u_i$.

These evolve according to the five coupled Equations in the absence of the Hubble expansion. The proof is given; the following is a summary of the equations:

$$\left[\frac{\partial_t^2 - c_\varphi^2\partial_i^2 - (1-c_\varphi^2)(\partial_i æ_i)\partial_t}{2N^2\Lambda_0}\right]\varphi - \frac{(\partial_t æ_i - \partial_i \Phi)}{\Lambda_0}\varphi$$

$$+F'\varphi = 0,$$

$$\partial_t\left[\lambda(\partial_t æ_i - \partial_i\Phi)\right] + \partial_i\left[\frac{(\partial_i æ_i + 3\partial_i\Phi)(-c_2)}{2}\right]$$

$$+\frac{1-c_\varphi^2}{2N^2}\partial_t\varphi\partial_i\varphi = 0,$$

$$\partial_i\left[\lambda\partial_i æ_i + (1-\lambda)\partial_i\Phi\right] = +4\pi G\rho^{bary},$$

where the time-derivative terms are all moved to the left-hand side, the summation over dummy index i = x, y, z is implicit, $\sqrt{\Lambda_0} \sim 10^{-10}\,\mathrm{ms}^{-2}, c\varphi \sim 1$, and N are constants, and we used $\varphi = \sqrt{\lambda}, \eta = -\partial_i æ_{i,} c_2 = -\frac{2}{3\lambda}$ and $c_4 = 2\lambda$.

These equations, when supplemented by the continuity equation and the three momentum equations for the normal baryonic fluid described by ρ^{bary} and v_i with a certain equation of state $P^{bary}(\rho^{bary})$, completely specify the dynamics.

The dark fluid has two types of deviations from MOND in general:

- The dark fluid has a generic scalar field vector oscillation and damping timescales $\sim O(v^{-1}) \sim O(\eta^{-1}) \sim O(N)$ times the orbital crossing time roughly unless the system is hotter than $c/\sqrt{N}$ or the forces are in resonance. A very fast damping would mean an almost instantaneous relation between gravity and the scalar field $1-\varphi^2$, as the μ_M in the classical Bekenstein & Milgrom modified gravity interpretation of MOND. A slow damping would mean a history-dependent relation, reminiscent of Milgrom's modified inertia interpretation of MOND: the dark fluid adds a dynamically varying inertia around the baryons that it surrounds. A possible test could be in galaxies with rotating bar(s), where there could be a phase lag between the bar and the effective DM. This has intriguing consequences to the bar's pattern speed because of nontrivial corrections to the MOND pictures of dynamical friction; the properties of the dark fluid are in between that of real particle dark halo and that naively expected from MOND.
- The dark fluid has a pressure, controlled by a propagation speed c_φ, where the speed of light is unity here, and the dark fluid can be made cold by $c^2\varphi \sim$ o, or hot by $c^2\varphi \sim 1$, or superluminal by $c^2\varphi \geq 1$. The φ would no longer be a function of the local gravity at $\mathbf{r}$ (as in MOND), rather it is a weighted average of a volume of all points $\mathbf{r}_1$ by a Yukawa-type screening function $\exp\left(-\frac{v|\mathbf{r}_1 - \mathbf{r}|}{c_\varphi}\right)$, where Screening length $= c_\varphi v^{-1} \sim c_\varphi \frac{10^8}{N} \times 300$ lt yr.
Note this spatial correction to MOND can exist even in static systems; even a small pressure term with $c^2\varphi \neq 0$ might smooth out MOND effects on small-scale structures (wide binaries, star clusters, dwarf galaxies), where the wave number $|\mathbf{k}|^2$ is much bigger than in galaxy clusters.

> The screening length can be set at ~100 pc for either an $N \sim 10^8$, $c_\varphi \sim 3\times10^5\,\text{km s}^{-1}$ hot dark fluid or an $N \sim 10^4$ and $c_\varphi \sim 30\,\text{km s}^{-1}$ cold dark fluid. This scale, 100 pc, is a scale dividing dense star clusters and fluffy dwarf galaxies. Observational DM effects are only seen in the universe on scales larger than 100 pc. It has been challenging for MOND to explain this observed scale.

We find a framework of dark fluid theories where MOND corresponds to a special choice of potentials or mass for the vector field. The dark fluid can run cold or hot depending on the sound speed c_φ (which could even be a running function of the vector field). These theories degenerate into scalar field theories for DE effects in the Hubble expansion. It is possible to create an exact w = −1 DE effect. The scale $a_0 = \sqrt{\Lambda_0}$ in MOND in equilibrium spiral galaxies derives its physics from the amplitude of the DE Λ_0. MOND or DM effects are hence indications of a non-uniform DE fluid described generally by a vector field $\varphi Æ^a$. For non-equilibrium systems like the Bullet clusters or galaxies with satellites, the properties of the dark fluid do not follow exactly the usual expectations of MOND or cold/hot DM, but (not so surprisingly) in between.

A Simple Lagrangian for the Dark Fluid

We propose a Lagrangian $\mathcal{L}$ containing a scalar φ and a time-like unit vector Æa, and the metric tensor g_{ab} as,

$$\mathcal{L} = R - \frac{c_\varphi^2}{N^2}\left(\nabla_c\varphi\nabla^c\varphi\right) + 2\Lambda_{\text{int}} + \mathcal{L}_Æ + I\!L = 0$$

Where, $N, c^2\varphi$ are various coupling constants or functions of $\lambda = \varphi^2$, R, is the Ricci scalar, and Λ_{int} takes the role of the scalar potential or interaction or the mass term:

$$\Lambda_{int}(\varphi, K_4) = \Lambda_0 F(\varphi^2) + \varphi^2 K_4, \quad K_4 \equiv \nabla_{\|}Æ_c\nabla_{\|}Æ^c,$$

where, $\Lambda_0 \equiv \left(1.2\times10^{-10}\,m\,s^{-1}\right)^2$ and K_4 are important for creating the MOND effect, and $K_4 = 0$ for a uniform cosmology; the notation $\nabla_{\|} \equiv Æ^a\nabla_a$ is a derivative parallel to the local time direction, and $\nabla^a = g^{ac}\nabla_c$, $Æ_a = g_{ac}Æ^c$. The term Λ_{int} takes on several roles: it gives an effective potential or mass for the scalar field, and allows the mass or sound speed to vary with the interaction of φ and Æ; to achieve the cold dark matter (CDM) effect during structure formation would require $\varphi^{-2}\Lambda_{\text{int}} \to \Lambda_0\varphi^{-2}F \to cst$ at high redshift when $\varphi \to \infty$, i.e., a nearly constant scale-free mass. To achieve the cosmological constant effect would require that $\Lambda_{\text{int}} \to \Lambda_0 F \to cst$ in the future when $\varphi \to 0$.

Other interaction terms are collected in:

$$\begin{aligned} I\!L &= e_0\varphi^2 R + \left[c_1\nabla_a Æ_c + c_3\nabla_c Æ_a\right] \\ &\quad -\left(e_1 Æ_a g_{bc} + e_2 Æ_b g_{ac}\right)\nabla^b\varphi Æ^c, \end{aligned}$$

$$\mathcal{L}_Æ = \left[c_2\left(\nabla_a Æ^a\right)^2 - \frac{1+c_\varphi^2}{N^2}\left(\nabla_{\|}\varphi\right)^2\right] + \left(g_{ab}Æ^a Æ^b - 1\right)L^*,$$

Where, L* is the Lagrangian multiplier.

In an effective sense, we can pack our $Æ^a$ and φ fields as one field: our Lagrangian can be cast in terms of a vector field of dynamics norm $Z^a \equiv Æ^a\varphi$ where one can pack up the 3+1 degrees of freedom into a vector of 4 degrees of freedom in the same style as in Halle et al.

For a first study, we shall set,

$$I\!\!L = 0$$

Hereafter by setting the coefficients c_1, c_3, e_0, e_1, e_2 (which are generally functions of φ) to zero. Specifically, we choose to set $e_0(\varphi) = 0$ to eliminate f(R)-gravity-like coupling and set the functions $e_1(\varphi) = e_2(\varphi) = 0$ to eliminate kinetic cross-coupling like $\nabla Æ \nabla \varphi$, and $c_1 = c_3 = 0$ to eliminate the spin-1 mode gravitational wave. This is done partly to simplify the dynamics, partly because any complicated couplings can be dangerous theoretically; these coupling terms serve as an effective potential for the scalar φ, and this dynamically changing potential does not appear to have a well-defined lower bound to safeguard the causal structure. It is tempting to set $c_2 = 0 = 1 - c^2\varphi$ for simplicity; however, we will not do so yet since these terms in $\mathcal{L}_Æ$ could be desirable for a theory to be healthy.

Parameters for MOND-Inspired Subset of Dark Fluid Theories

A theory is built by minimizing the action $S = -(16\pi G)^{-1} \int dx^4 \sqrt{-g}\,(\mathcal{L}_m + \mathcal{L})$ once the Lagrangian density $\mathcal{L}$ is specified. Our Lagrangian $\mathcal{L}$ has three dynamical fields: the scalar field λ (or equivalently φ, the Æther field Æa, and the metric field g^{ab}. Now minimizing the action S with respect to variations of the three fields will lead to the scalar field EOM, Æther field EOM, and the modified energy–momentum tensor. Variations with the non-dynamical Lagrange multiplier L^* gives the unit-vector constraint for the Æ field. The general results are more tedious and are presented elsewhere. To illustrate the physics and stay away from models with obvious pathological dynamics, we consider only a specific choice of functions to select a MOND-related subset of dark fluid theories. Starting from the Lagrangian given in Equation $\mathcal{L} = R - \frac{c_\varphi^2}{N^2}(\nabla_c \varphi \nabla^c \varphi) + 2\Lambda_{\text{int}} + \mathcal{L}_Æ + I\!\!L = 0$, we select models with $I\!\!L = 0$ and select the functions in $\mathcal{L}_\varphi$ and $\mathcal{L}_Æ$ as follows:

The dimensionless function $F(\lambda)$ has the meaning of the potential of the scalar field $\lambda \equiv \varphi^2$. the coupling constants $c^2\varphi$ and c_2 are of order unity, and will be shown to be related to the sound speeds of the dark fluid.

1. Our choice for the dark fluid potential with a constant energy scale Λ_0 is:

   $$\Lambda_0 F(\varphi^2) = \frac{k^3\Lambda_0}{3} + \frac{k^3\Lambda_0}{3}\left(|\lambda_\pm|^{1/k} - 1\right)^3, \quad k = 3,$$

   where we shall adopt $\lambda_\pm \approx \lambda = \varphi^2$ essentially up to a small correction. The reason to choose k = 3 is such that we make the potential $\Lambda_0 F \to 9\Lambda_0\varphi^2$ to be harmonic at large φ to mimic CDM. In galaxies we can neglect ε; so $\lambda_\pm \approx \lambda \equiv \varphi^2$ and $F = 9(\varphi^{2/3} - 1)^3 + 9 = 9(\varphi^2 - 3\varphi^{4/3} + 3\varphi^{2/3})$. An important asymptotic property of the functions F and F' (a prime ' always means a derivative $\frac{d}{d\lambda}$ is that,

   $$F' \equiv \frac{d}{d\lambda}F \sim \left(1 - \lambda^{-1/3}\right)^2, \text{if } \lambda \equiv \varphi^2 \to 1,$$

   $$\propto \left(\lambda - \varepsilon^2\right)^{-1/2}, \text{if } \lambda \equiv \varphi^2 \to +\varepsilon^2 \sim 10^{-7},$$

   $$\sim 0 \text{ if } \lambda \equiv \varphi^2 \to \infty.$$

We shall show that this property describes a non-uniform (DE) fluid which gives the MOND-like (DM) effects in galaxies, and Newtonian-like effects in the solar system.

2. We set the scalar field sound speed as a finite positive constant, i.e., $\infty > c_\varphi > 0,$

   and we shall treat N as constants too. In general, theories with finite positive propagation speeds have a finite causality cone, hence are well behaved. The opening angle of the cone of propagation can be wider than that of photons.

3. We set the equivalent part of Jacobson's unit vector field Lagrangian to be $c_1K_1 + c_3K_3 + c_4K_4 + c_2K_2 = 0\times K_1 + 0\times K_3 + 2\lambda K_4 + c_2K_2$, where, e.g., c_2K_2 corresponds to the second term in $\mathcal{L}_{Æ}$. This choice $c_1 = c_3 = 0$ in the Lagrangian kills spin-1 mode waves of the vector field, and guarantees that the normal gravitational wave in the tensor mode will propagate with the normal speed (of light). This might not be necessary, but it simplifies the analysis of parameterized post-Newtonian (PPN) parameters in the solar system and the sound speed of the vector field. One can use the analysis by Foster & Jacobson to predict the sound speed square of the spin-0 mode of Aether fluctuation, $c_0^2 \equiv \frac{(c_2+c_1+c_3)}{(c_4+c_1)(1-c_1-c_3)}\frac{(2-c_4-c_1)}{(2+c_1+3c_2+c_3)} = \frac{(1-\lambda)c_2}{(2+3c_2)\lambda}$. Among the many choices of c_2 to guarantee that $c^2{}_0$ and c_2 are smooth functions for all values of λ, the simplest choice is such that,

   $-\frac{c_2}{2} = \frac{1}{3\lambda} \rightarrow c_0^2 = \frac{1}{3\lambda}.$

   The important thing is that c20 is a finite real positive number given by $c_0^2 = \frac{1}{3\lambda} > 0$. Interestingly, in the solar system, where the scalar field λ is expected to settle to a very small equilibrium value $\lambda \sim \varepsilon^2 \sim O(10^{-7})$ for our choice of the penalizing scalar field potential F(λ), the spin-0 mode of the vector field propagates with a finite superluminal speed, $c_0 = (3\varepsilon^2)^{-1/2}\, Gt1,$ avoiding the Cherenkov radiations constraint in the solar system. All PPN parameters are expect to be equal to that of GR in the solar system as well; the PPN parameters $\alpha_1 = -8\lambda$, $\alpha_2 = (3\lambda - 1)\lambda$ are non-zero, but can be made small enough to fit current limits $|\alpha_1| < 10^{-4}$ and $|\alpha_2| < 4\times10^{-7}$.

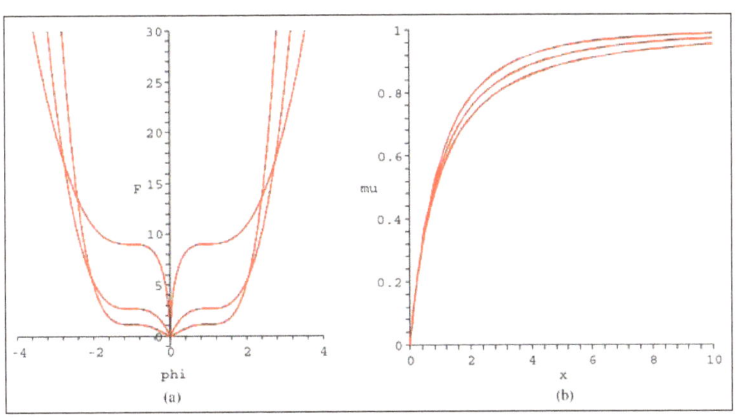

In the figure above Panel (a): $F(\varphi^2)$ vs. φ for k = 3, 3/2, 2 (top to bottom); note the symmetric potential with the global minimum F = 0 at φ = 0, and the plateau at λ = 1. Panel (b) shows our

function $\mu \equiv 1 - \varphi^2$ as function of $x \equiv |\nabla\Phi| / \sqrt{\Lambda_0}$ for the models with k = 3, 3/2, 2 (top to down), adopting $\sqrt{\Lambda_0} \to a_0$.

Dark Matter

Dark matter is a component of the universe whose presence is discerned from its gravitational attraction rather than its luminosity. Dark matter makes up 30.1 per cent of the matter-energy composition of the universe; the rest is dark energy (69.4 per cent) and "ordinary" visible matter (0.5 per cent).

Originally known as the "missing mass," dark matter's existence was first inferred by Swiss American astronomer Fritz Zwicky, who in 1933 discovered that the mass of all the stars in the Coma cluster of galaxies provided only about 1 per cent of the mass needed to keep the galaxies from escaping the cluster's gravitational pull. The reality of this missing mass remained in question for decades, until the 1970s when American astronomers Vera Rubin and W. Kent Ford confirmed its existence by the observation of a similar phenomenon: the mass of the stars visible within a typical galaxy is only about 10 per cent of that required to keep those stars orbiting the galaxy's centre. In general, the speed with which stars orbit the centre of their galaxy is independent of their separation from the centre; indeed, orbital velocity is either constant or increases slightly with distance rather than dropping off as expected. To account for this, the mass of the galaxy within the orbit of the stars must increase linearly with the distance of the stars from the galaxy's centre. However, no light is seen from this inner mass—hence the name "dark matter."

Since the confirmation of dark matter's existence, a preponderance of dark matter in galaxies and clusters of galaxies has been discerned through the phenomenon of gravitational lensing—matter acting as a lens by bending space and distorting the passage of background light. The presence of this missing matter in the centres of galaxies and clusters of galaxies has also been inferred from the motion and heat of gas that gives rise to observed X-rays. For example, the Chandra X-ray Observatory has observed in the Bullet cluster, which consists of two merging galaxy clusters, that the hot gas (ordinary visible matter) is slowed by the drag effect of one cluster passing through the other. The mass of the clusters, however, is not affected, indicating that most of the mass consists of dark matter.

Gravitational lens.

In this picture a galactic cluster, about five billion light-years away, produces a tremendous gravitational field that "bends" light around it. This lens produces multiple copies of a blue galaxy about twice as distant. Four images are visible in a circle surrounding the lens; a fifth is visible near the centre of the picture, which was taken by the Hubble Space Telescope.

Composite image showing the galaxy cluster 1E0657-56, the Bullet cluster.

Matter is 30.6 per cent of the universe's matter-energy composition. Only 0.5 per cent is in the mass of stars and 0.03 per cent of that matter is in the form of elements heavier than hydrogen. The rest is dark matter. Two varieties of dark matter have been found to exist. The first variety is about 4.5 per cent of the universe and is made of the familiar baryons (i.e., protons, neutrons, and atomic nuclei), which also make up the luminous stars and galaxies. Most of this baryonic dark matter is expected to exist in the form of gas in and between the galaxies. This baryonic, or ordinary, component of dark matter has been determined by measuring the abundance of elements heavier than hydrogen that were created in the first few minutes after the big bang occurred 13.8 billion years ago.

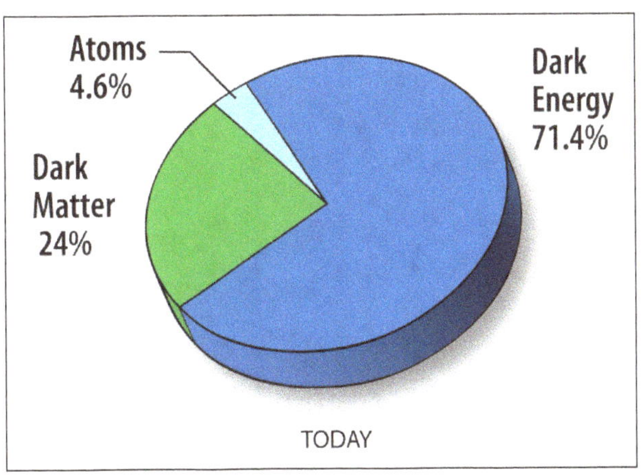

The dark matter that comprises the other 26.1 per cent of the universe's matter is in an unfamiliar, nonbaryonic form. The rate at which galaxies and large structures composed of galaxies coalesced from density fluctuations in the early universe indicates that the nonbaryonic dark matter is relatively "cold," or "non-relativistic," meaning that the backbones of galaxies and clusters of galaxies are made of heavy, slow-moving particles. The absence of light from these particles also indicates that they are electromagnetically neutral. These properties give rise to the particles' common name, weakly interacting massive particles (WIMPs). The precise nature of these particles is not currently known, and they are not predicted by the standard model of particle physics. However, a number of possible extensions to the standard model such as super-symmetric theories predict hypothetical elementary particles such as axions or neutralinos that may be the undetected WIMPs.

Extraordinary efforts are under way to detect and measure the properties of these unseen WIMPs, either by witnessing their impact in a laboratory detector or by observing their annihilations after they collide with each other. There is also some expectation that their presence and mass may be inferred from experiments at new particle accelerators such as the Large Hadron Collider.

As an alternative to dark matter, modifications to gravity have been proposed to explain the apparent presence of "missing matter." These modifications suggest that the attractive force exerted by ordinary matter may be enhanced in conditions that occur only on galactic scales. However, most of the proposals are unsatisfactory on theoretical grounds as they provide little or no explanation for the modification of gravity. These theories are also unable to explain the observations of dark matter physically separated from ordinary matter in the Bullet cluster. This separation demonstrates that dark matter is a physical reality and is distinguishable from ordinary matter.

Lambda CDM Model

The Lambda Cold Dark Matter (or LCDM) model is the latest incarnation of our understanding about the origin of the Cosmos. It represents an improvement of the big bang theory by positing most of the physical substance in the Universe consists of a material called dark matter.

Although it cannot be detected by current instrumentation, cosmologists believe dark matter is comprised of cold slow moving particles that do not emit electromagnetic radiation or scatter light, thus they also appear dark. However, the gravitational effect of dark matter can be observed on visible material, such as galaxies and observations of background radiation.

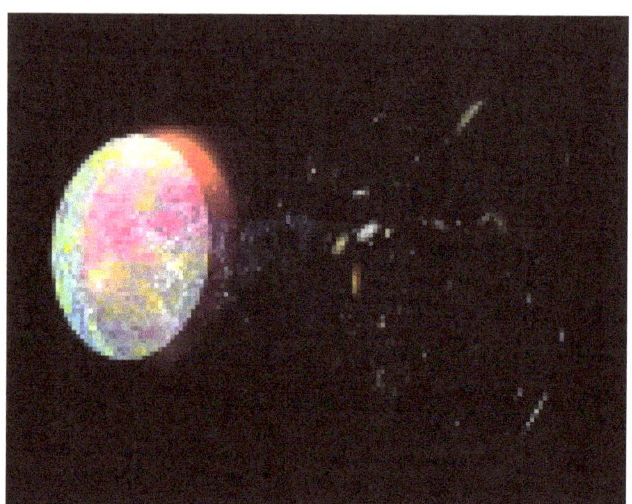

Beginner guide to the LCDM Universe.

Is it possible for distant galaxies to be moving away from us faster than the speed of light? Watch an accurate picture of the Universe based on the Lambda Cold Dark Matter model, the best cosmological model today.

The Lambda in the theory's name accounts for the presence of dark energy, a hypothetical force that appears to be accelerating the expansion of the Universe. Today, it's also referenced as the standard cosmological model.

According to the LCDM theory, the Universe was intensely hot, remarkably smooth and essentially homogeneous immediately following the Big Bang. However, small fluctuations in density, less than one part in a hundred thousand, began to appear and grow. As the universe cooled, clumps of dark matter began to condense and within them gas molecules formed. At this point the universe was almost exclusively composed of hydrogen, helium, and dark matter. This is the period when the cosmic microwave background radiation was emitted.

The gas and dark matter were gravitationally attracted to the areas of higher density and formed halos that represented the seeds for the first galaxies. As the halos became more massive, they began to collapse under their own weight and became proto-galaxies. Soon after, the hydrogen and helium gas within the halos began to make the first stars. Then, over time, the halos merged to form larger and larger galaxies.

Clumpy early universe.

Small dark matter halos can be seen merging in this computer simulation of the early Universe based on the Lambda Cold Dark Matter theory. Overtime, the mergers produce a proto-galaxy.

Computer simulations have enabled theorists to follow the evolution of matter in the Universe from a short time after the Big Bang until recently. These simulations contend that galaxy formation in the presence of cold dark matter occurs hierarchically - the first galaxies to form are small dwarfs and these subsequently merge to form progressively larger star systems. Therefore, large galaxies like the Milky Way must have consumed a hundred or more small dwarf galaxies over time. However, many of them may be so thoroughly mixed with the stars of their parent galaxy they are no longer easily identifiable.

Rings and arcs aren't the only evidence of ancient satellite mergers. This figure simulates the merger of numerous companion galaxies and demonstrates that plumes, spears, spikes and shells that surround the primary galaxy are also possible.

Mergers Made in Heaven

The most spectacular manifestation of this process may be the coalescence of comparably sized galaxies in a process known as a major merger. These events often result in the destruction of the spiral pattern in both merging galaxies. Major mergers can also trigger star bursts. Such events have been relatively rare within the past few billion years and only a small percentage of large galaxies are involved in an ongoing major merger at any point in time.

However, minor mergers that involve the disruption of a dwarf satellite galaxy by a much more massive companion are expected to be significantly more common. According to the LCDM model, minor mergers should still be occurring today. Since the stellar disk of the larger partner is not destroyed during a minor merger, signs of recent or on-going events should be apparent around many spirals, the most common type of large galaxy.

Cosmic Bread Crumbs

Sagittarius Dwarf Galaxy.

The Milky Way's Sagittarius tidal stream can be seen extending from the dense 'core' of the Sagittarius dwarf, wrapping around the galaxy, and descending through the Sun's position.

The ebb and flow of the tides are created when the Moon attracts our planet and its oceans differentially. Similarly, a large spiral deforms an orbiting satellite galaxy by exerting a stronger pull on one side than the other. In the process, some of the satellite's stars are removed like a trail of bread crumbs that leaves a fossil record in the form of a stellar tidal structure. During their merger, stars from a satellite galaxy can be pulled into long streams, deposited into vast shells of debris or swept into enormous umbrella shaped structures that surround the parent galaxy and remain detectable for several billion years like a gigantic cosmic relic.

Evidence of the first stellar stream was discovered in the Milky Way galaxy during the 90s. Known as the Sagittarius Dwarf Elliptical, this small satellite galaxy is orbiting on a perpendicular path to the Milky Way's broad stellar plane causing it to pass through our galaxy. With each pass through the disk, stars are being removed forming a thin stream.

Since the discovery of the Sagittarius dwarf galaxy, over 15 star streams have been identified in the halo our home galaxy and 4 have been discovered in the Andromeda galaxy, our nearest galactic neighbor. However, evidence of tidal streams beyond the Local Group has been mostly anecdotal until recently.

Beyond the Local Group

For the past six years, an international team of professional and amateur astronomers searching for stellar streams around eight near-by spirals beyond the Local Group by analyzing ultra-deep images produced with modest-sized, commercially available instruments. Their efforts led to the discovery of six extensive stellar structures surrounding several of the galaxies surveyed. These previously undetected features were interpreted as debris from tidally disrupted satellites. In addition, their research confirmed and clarified several enormous stellar features that had been previously reported but never interpreted as fossil evidence of minor mergers.

Galaxy in a bubble.

NGC3521 is located about 35 million light years in the distance toward the northern constellation of Leo. This new deep image depicts multiple debris shells that evidence previous mergers with one or more satellite galaxies.

The set of galaxies exhibited unexpected, highly diverse characteristics such as great circular features resembling the Milky Way's Sagittarius stream, remote shells and giant clouds of tidal debris as well as enormous jet-like features emerging from galactic disks. Together with the remains of already disrupted companions, the observations also captured surviving satellites caught in the act of tidal disruption.

Earlier in 2011, a new image of NGC 3521 was completed that improves on the data originally collected for the survey. Located 35 million light years from our planet towards the northern constellation of Leo, this star system has been classically categorized as flocculent galaxy due to the enormous amount of material partially obscuring its spiral structure.

However, the new ultra-deep image reveals evidence of one or more previous mergers with dwarf galaxies that left discernible substructures, such as an almost spherical cloud of debris visible on its eastern side and a large, elongated cloud to the west. Both represent debris shells belonging to an umbrella-like structure similar to the one seen in images of NGC 4651. But, their looser appearance suggests they were accreted much farther in the past. Additionally, the galaxy is enveloped a bubble of multiple debris shells that may represent further evidence of ancient mergers.

The Umbrella Galaxy, NGC4651, is located in the constellation of Coma Berenices and situated about 35 million light years from our planet. This new view displays evidence of previous mergers with one or more satellite galaxies.

The shell of stellar debris apparently pierced by a narrow tidal 'spear' is consistent with predictions based on the standard model. Other flocculent galaxies have also been shown to have stellar stream remnants, such as NGC 5055 (M63) leading some to speculate that this phenomenon may be linked to previous minor mergers.

A comparison with computer simulations confirmed the extraordinary variety of structures detected by the Delgado team. The existence of the tidal features around distant galaxies that appear to be normal, from all other aspects, and their match to the LCDM simulations constituted new evidence that the standard model also applies to a distant galaxies similar to the Milky Way.

Like cosmic archaeologists, astronomers excavate ancient light to uncover the truth about the birth and development of galaxies. Stellar streams, relics of the hierarchical merger process predicted by the LCDM theory, were first identified in the Milky Way and other galaxies in the Local Group. Now, similar structures have been seen in much more distant galaxies. Combined, these observations support the current best theory explaining how the Universe, and everything within it, originated and evolved to the present day.

However, the book of scientific knowledge is written on loose leaf pages that are revised, re-sorted and sometimes removed over time when new information is dug up. Therefore, unlike earth-bound fossils, nothing cosmological should be considered etched in stone because there's always something new coming to light just over our horizon.

Cosmological Constant

In the context of cosmology the cosmological constant is a homogeneous energy density that causes the expansion of the universe to accelerate. Originally proposed early in the development of general relativity in order to allow a static universe solution it was subsequently abandoned when the universe was found to be expanding. Now the cosmological constant is invoked to explain the observed acceleration of the expansion of the universe. The cosmological constant is the simplest realization of dark energy, which the more generic name is given to the unknown cause of the acceleration of the universe. Its existence is also predicted by quantum physics, where it enters as a form of vacuum energy, although the magnitude predicted by quantum theory does not match that observed in cosmology.

The cosmological constant was originally introduced by Einstein in 1917 as a repulsive force required keeping the Universe in static equilibrium. In modern cosmology it is the leading candidate for dark energy, the cause of the acceleration of the expansion of the universe.

The cosmological constant first appeared in a 1917 paper by Einstein entitled "Cosmological Considerations in the General Theory of Relativity", in which he motivates its introduction into the field theory of general relativity by the need to stabilize the universe against the attractive effect of gravity:

"The term is necessary only for the purpose of making possible a quasi-static distribution of matter, as required by the fact of the small velocities of the stars".

At the time, observations of our universe were limited primarily to stars in our own galaxy, so there was indeed observational evidence justifying the assumption that the universe was static.

Einstein's goal was to obtain a Universe that satisfied Mach's principle of the relativity of inertia, and construct a cosmology that was finite, yet stable against gravitational collapse. The attempt proved futile, as shortly thereafter de Sitter demonstrated an empty universe solution to Einstein's equations (allowing inertia relative to space empty of matter) and Friedmann derived solutions to Einstein's equations that corresponded to an expanding universe. These results could be considered a prediction that the universe must be expanding or contracting, a remarkable implication of general relativity that was later borne out by observation. When Hubble observationally discovered the expansion of the universe Einstein finally abandoned the cosmological constant completely.

In the intervening years the cosmological constant came in and out of vogue as new observational results repeatedly seemed to require it, but then were explained in other ways. As of the early 1990s there were tantalizing hints that the cosmological constant might again be needed. The universe appeared to be younger than the oldest stars it contained, a feature that was remedied if the universe was currently in an accelerating phase. Number counts of galaxies indicated that the volume contained within a solid angle at high redshift was larger than expected in a decelerating universe. Theoretical arguments from inflation and later observational results from the cosmic microwave background radiation indicated that the universe should be flat, but observations of large scale structure indicated that the matter density was inadequate to achieve this -- vacuum energy could make up the shortfall.

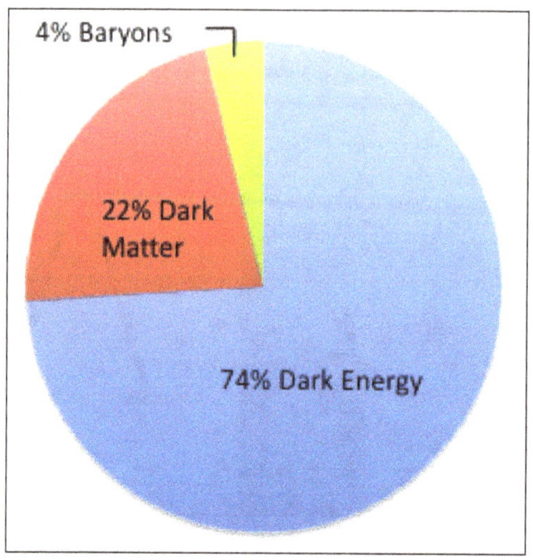

The mass-energy composition of the Universe. The cosmological constant is one possible form of dark energy, which is believed to be behind the acceleration of the expansion Universe.

This set the stage for the discovery of the accelerating universe by two teams in 1998/1999. The High-Z supernova team and the Supernova Cosmology project both discovered that high-redshift supernovae were fainter than expected for a decelerating universe and that the difference could be explained if there was a cosmological constant of just the right magnitude needed to make the universe flat.

This was a dramatic convergence of observation and theory. Since then increasingly accurate probes have confirmed to high precision the need for dark energy, but the nature of the dark energy is now the issue being investigated. As of 2010 the measured properties of dark energy remain consistent with those of a cosmological constant. However, massive observing efforts are underway to test whether this is the correct explanation for the acceleration or whether some other sort of dark energy, perhaps one that changes with time or one that is motivated by some form of quantum gravity, is needed to explain the acceleration.

The Physics of the Cosmological Constant

To explore more deeply the nature of the Universe, we must use the mathematical language in Einstein's general relativity to relate the geometry of space-time (expressed by the metric tensor, $g_{\mu\nu}$) to the energy content of the universe, (expressed by the energy-momentum tensor, $T_{\mu\nu}$).

Einstein Field Equations

Arguably, one of Einstein's most significant discoveries was that the distribution of energy determines the geometry of space-time, which is encoded in his field equation,

$$R_{\mu\nu} - \frac{1}{2} R g_{\mu\nu} = 8\pi G T_{\mu\nu}$$

where G is the gravitational constant. Although this is the simplest form of the equations the freedom remains to add a constant term. This "cosmological constant" was what Einstein added in order to achieve a static universe, and it is given the symbol Λ,

$$R_{\mu\nu} - \frac{1}{2} R g_{\mu\nu} + \Lambda g_{\mu\nu} = 8\pi G T_{\mu\nu}$$

When Λ is positive it acts as a repulsive force.

Vacuum Energy

Equation of State (w)	
Radiation	1/3
Matter (pressureless)	0
Curvature	-1/3
Cosmological Constant	-1
....	
Matter (general)	0<w<1/3
Quintessence	-1<w<1/3

The equation of state, w, describes the relationship between pressure and density in a material according to $w = p / \rho$. Here are some examples of the equation of state for common fluids. When matter is at rest (pressure less dust) it has w = 0, but as it picks up velocity (v) its equation of state increases until $w \to 1/3$ as $v \to c$.

Vacuum energy arises naturally in quantum mechanics due to the uncertainty principle. In particle physics the vacuum refers to the ground state of the theory - the lowest energy configuration. The uncertainty principle does not allow states of exactly zero energy, even in vacuum (virtual particles are created). Since in general relativity all forms of energy gravitate, this ground state vacuum energy impacts the dynamics of the expansion of the universe.

Vacuum energy should not have any dissipative processes such as heat conduction or viscosity, so it should take the form of a perfect fluid,

$$T_{\mu\nu} = (\rho + p)U_\mu U_\nu + p g_{\mu\nu}.$$

In order to maintain Lorentz invariance, vacuum energy should also have no preferred direction. Therefore the first term in the perfect fluid energy tensor must be zero, requiring,

$$p^{vac} = -\rho^{vac},$$

which corresponds to an equation of state $w^{vac} = p^{vac} / \rho^{vac} = -1$, and results in an energy-momentum tensor for vacuum energy,

$$T_{\mu\nu}^{vac} = p^{vac} g^{\mu\nu} = -\rho^{vac} g_{\mu\nu}.$$

Equivalence of Cosmological Constant and Vacuum Energy

We can split the energy-momentum tensor into a term describing the matter and energy, and a term describing the vacuum, $T_{\mu\nu} = T_{\mu\nu}^{matter} + T_{\mu\nu}^{vac}$. Einstein's equation including vacuum energy becomes,

$$R_{\mu\nu} - \frac{1}{2} R g_{\mu\nu} = 8\pi G\left(T_{\mu\nu}^{matter} - \rho_{vac} g_{\mu\nu}\right).$$

Recall that the cosmological constant enters Einstein's equation in the form,

$$R_{\mu\nu} - \frac{1}{2} R g_{\mu\nu} + \Lambda g_{\mu\nu} = 8\pi G T_{\mu\nu}.$$

So vacuum energy and the cosmological constant have identical behavior in general relativity, as long as the vacuum energy density is identified with,

$$\rho^{vac} = \frac{\Lambda}{8\pi G}.$$

Cosmology

In a homogeneous, isotropic universe the geometry is defined by the Friedmann-Lemaître-Robertson-Walker metric (FLRW metric) and the dynamics of the universe are governed by the Friedmann equations (Friedmann equations). The dynamics are driven by the energy content of the universe and the equation of state of the components that make up the energy density. The equation of state relates density ρ to pressure p according to $w=p/\rho$. The cosmological constant enters these equations in the following way, where a is the scale factor of the universe normalized to 1 at the present day, $H = \dot{a}/a$ is Hubble's constant (an overdot represents differentiation with respect to time), and k is the curvature of the universe given by +1, 0, and −1 for positive, flat, and negative curvature respectively,

$$H^2 = \frac{8\pi G}{3}\rho - \frac{k}{a^2} + \frac{\Lambda}{3},$$

$$\frac{\ddot{a}}{a} = -\frac{4\pi G}{3}(\rho + 3p) + \frac{\Lambda}{3}.$$

These equations are more concisely written by considering both the cosmological constant and curvature as forms of energy density ($\rho_\Lambda = \Lambda / 8\pi G$ and $\rho_k = -3k / 8\pi G a^2$). Then Equations $H^2 = \frac{8\pi G}{3}\rho - \frac{k}{a^2} + \frac{\Lambda}{3}$ and $\frac{\ddot{a}}{a} = -\frac{4\pi G}{3}(\rho + 3p) + \frac{\Lambda}{3}$ become,

$$H^2 = \frac{8\pi G}{3}\sum_i \rho_i,$$

$$\frac{\ddot{a}}{a} = -\frac{4\pi G}{3}\sum_i (\rho_i + 3p_i).$$

The different components have different equations of state, w_i, which determines how their density changes with the expansion of the universe:

$$\rho_i = \rho_{i0} a^{-3(1+w_i)}$$

Pressure less matter has ω = 0, radiation has ω = 1/3, curvature has an effective ω = −1/3, cosmological constant has w = −1. (We have used $w_\Lambda = -1$, which implies $\rho_\Lambda + 3p_\Lambda = -2\rho_\Lambda$ in deriving Equation. $\frac{\ddot{a}}{a} = -\frac{4\pi G}{3}\sum_i (\rho_i + 3p_i)$.

The current energy density of each component, ρ_{i0}, is often represented as a fraction of the critical density, $\rho_c = 3H_0^2 / 8\pi G$, which is the energy density required to close the universe (also calculated at the present day). Denoting this $\Omega_i = \rho_{i0} / \rho_c$ and using Eq. $\rho_i = \rho_{i0} a^{-3(1+w_i)}$ allows us to write,

$$H^2 = H_0^2 \sum_i \frac{\rho_i}{\rho_c} = H_0^2 \sum_i \Omega_i a^{-3(1+w_i)}$$

For pressure to do work there needs to be a pressure gradient - a relatively high pressure region next to a relatively low pressure region -- that will then cause movement from high pressure to low. In a homogeneous universe there are no pressure gradients, so a positive pressure does no work and has no expanding effect (there are no low-pressure regions for it to push matter into). On the contrary, in general relativity all forms of energy gravitate so pressure effectively pulls, strengthening the attractive force of gravity (thus the factor of p in Equation. $\frac{\ddot{a}}{a} = -\frac{4\pi G}{3}(\rho + 3p) + \frac{\Lambda}{3}$., which does not appear in Newtonian gravity). The cosmological constant has negative pressure, w = −1, so its general relativistic contribution counteracts the normal force of gravity and provides an outwards acceleration.

Observational Evidence

Observational evidence for the accelerating universe is now very strong, with many different experiments covering vastly different timescales, length scales, and physical processes, all supporting the standard ΛCDM cosmological model, in which the universe is flat with an energy density made up of about 4% baryonic matter, 23% dark matter, and 73% cosmological constant.

The critical observational result that brought the cosmological constant into its modern prominence was the discovery that distant type Ia supernovae ($0 < z < 1$), used as standard candles, were fainter than expected in a decelerating universe. Since then many groups have confirmed this result with more supernovae and over a larger range of redshifts. Of particular importance are the observations that extremely high redshift ($z > 1$) supernovae are brighter than expected, which is the observational signature that is expected from a period of deceleration preceding our current period of acceleration. These higher-redshift observations of brighter-than-expected supernovae protect us against any systematic effects that would dim supernovae for reasons other than acceleration.

Prior to the 1998 release of the supernova results there were already several lines of evidence that paved the way for the relatively rapid acceptance of the supernova evidence for the acceleration of the universe. Three in particular included:

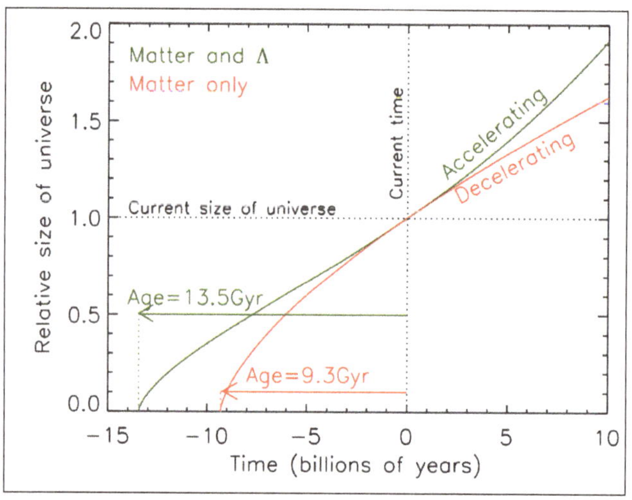

The relative size of the universe as a function of time for a flat universe made entirely of matter (red) and one made of 30% matter and 70% cosmological constant (green).

In both cases the zero of time corresponds to the present day, and that has been defined so that the slope matches the current expansion rate of the universe (Hubble's constant is taken to be 70 km/s/Mpc). Both types of universe would have initially decelerated, but the universe with the cosmological constant later switched and started accelerating. The cosmological constant universe is older because it took longer to reach its present rate of expansion (13.5 Gyr) than the matter-only universe (9.3 Gyr).

- The universe appeared younger than the oldest stars. Stellar evolution is well understood, and observations of stars in globular clusters and elsewhere indicate that the oldest stars are over 13 billion years old. We can compare this to the age of the universe by measuring the universe's rate of expansion today and tracing that back to the time of the big bang. If the universe had decelerated to its current speed then the age would be lower than if it had accelerated to its current speed. A flat universe made only of matter would only be about 9 billion years old -- a major problem given that this is several billion years younger than the oldest stars. On the other hand, a flat universe with 74% cosmological constant would be about 13.7 billion years old. Thus the observation that the universe is currently accelerating solved the age paradox.

- There were too many distant galaxies. Galaxy number counts had already been used widely in attempts to estimate the deceleration of the expansion of the universe. The volume of space between two redshifts differs depending on the expansion history of the universe (for a given solid angle). Using the number of galaxies between two redshifts as a measure of the volume of space, observers had measured that distant volumes seemed too large compared with the predictions of a decelerating universe. Either the luminosity of galaxies or the number of galaxies per unit volume was evolving with time in an unexpected way, or the volumes we were calculating were incorrect. An accelerating universe could explain the observations without invoking any strange galaxy evolution.
- The observed flatness of the universe despite insufficient matter. Using measurements of temperature fluctuations in the cosmic microwave background radiation (CMB) from when the universe was ~380,000 years old one can conclude that the Universe is spatially flat to within a few percent. By combining these data with accurate H_0 measurements and/or measurements of the matter density of the universe, it becomes clear that the matter in the Universe only contributes approximately 23% to the critical density. One way to account for the missing energy density would be to invoke a cosmological constant. As it turns out, the amount of cosmological constant needed to explain the acceleration observed in the supernova data, was just what was needed to also make the universe flat. Therefore the cosmological constant solved the apparent contradiction between the matter-density and CMB observations.

Unresolved Issues

Despite its success, the cosmological constant is not without problems.

Cosmological Constant Problem

The cosmological constant problem arises because; using naive naturalness arguments in quantum field theory, and one cannot explain why the observed cosmological constant is so small. Quantum mechanical calculations that sum the contributions from all vacuum modes below an ultraviolet cutoff at the Planck scale give a vacuum energy density of $\rho_\Lambda \sim 10^{112}\,\mathrm{erg/cm^3}$. This exceeds the cosmologically observed value of $\rho_\Lambda \sim 10^{-8}\,\mathrm{erg/cm^3}$ by about 120 orders of magnitude.

Coincidence Problem

The cosmological constant is not diluted as the universe expands, whereas the density of matter drops in inverse proportion to the volume. This means that there is only a fleeting moment of cosmological time during which the matter density will be of comparable magnitude to the vacuum energy density. Many argue that to be living in that moment is too unlikely to be coincidence. This has been called the coincidence problem, and has motivated theories beyond the cosmological constant with more general forms of dark energy that may change with time.

Dark Energy or Cosmological Constant

These unresolved issues have motivated the current observational effort to test whether the cosmological constant is a valid cause of the acceleration of the universe. Other theories, such as fledgling theories of quantum gravity (e.g. brane-motivated cosmologies), naturally produce dark energy candidates with properties different from the standard cosmological constant. Phenomenological theories such as quintessence have also been proposed, which have a time-varying value of dark energy. Although these models are designed partially to negate the coincidence problem by having dark energy solutions that can track the matter density, these then suffer a new fine-tuning problem as they introduce additional parameters whose values need to be fine-tuned to produce the evolution needed.

Anthropic Solutions

Many argue that the coincidence problem is most simply solved by anthropic considerations. That is, were the value of the cosmological constant much higher or lower than the observed value it would disrupt structure formation in the universe and humans would not exist. Although many argue against anthropic solutions on philosophical grounds in preference for solutions that invoke some deeper physical principles, anthropic arguments are gaining more prominence, especially in light of the emergence of the string landscape. It had been hoped that string theory would give a well-motivated fundamental explanation for the values of the constants of nature, but now it seems ours is only one of many possible solutions, which we find ourselves in by chance constrained by anthropic requirements.

Dark Gravity

Dark energy also encompasses the possibility that there is no additional energy density component to the universe, but rather that the equations of general relativity need revision. In this sense general relativity might be a limit of a more complete theory of gravity in the same way that Newtonian gravity is a low-energy limit of general relativity. This possibility is also known as dark gravity.

References

- The-Cosmological-Spacetime-45865140: researchgate.net, Retrieved 11, June 2020
- Galaxy-formation-and-evolution: cosmotography.com, , Retrieved 25, june 2020
- Cosmological-constant: scholarpedia.org, Retrieved 19, March 2020
- Dark-matter: britannica.com, Retrieved 16, March 2020
- History-of-cosmic-structure-formation, Space-Science, Science-Exploration: esa.int, Retrieved 20, March 2020

3

Special and General Relativity

The theory of special relativity determines the relationship of space and time for objects moving in a straight line at a constant speed. The topics elaborated in this chapter are Lorentz transformations, tests of special relativity, general relativity, Einstein equation, etc. It discusses in detail the theories and methodologies related to special and general relativity.

Special Relativity

The theory of special relativity explains how space and time are linked for objects that are moving at a consistent speed in a straight line. One of its most famous aspects concerns objects moving at the speed of light.

Simply put, as an object approaches the speed of light, its mass becomes infinite and it is unable to go any faster than light travels. This cosmic speed limit has been a subject of much discussion in physics, and even in science fiction, as people think about how to travel across vast distances.

The theory of special relativity was developed by Albert Einstein in 1905, and it forms part of the basis of modern physics. After finishing his work in special relativity, Einstein spent a decade pondering what would happen if one introduced acceleration. This formed the basis of his general relativity, published in 1915.

Before Einstein, astronomers (for the most part) understood the universe in terms of three laws of motion presented by Isaac Newton in 1686. These three laws are:

- Objects in motion (or at rest) remain in motion (or at rest) unless an external force imposes change.
- Force is equal to the change in momentum per change of time. For a constant mass, force equals mass times acceleration.
- For every action, there is an equal and opposite reaction.

But there were cracks in the theory for decades before Einstein's arrival on the scene. In 1865, Scottish physicist James Clerk Maxwell demonstrated that light is a wave with both electrical and magnetic components, and established the speed of light (186,000 miles per second). Scientists supposed that the light had to be transmitted through some medium, which they called the ether. (We now know that no transmission medium is required, and that light in space moves in a vacuum.)

An unexpected result threw this into question. Physicist A.A. Michelson and chemist Edward Morley calculated how Earth's motion through this "ether" affected how the speed of light is measured, and found that the speed of light is the same no matter what Earth's motion is. This led to further musings on light's behaviour — and its incongruence with classical mechanics — by Austrian physicist Ernst Mach and French mathematician Henri Poincare.

Einstein began thinking of light's behaviour when he was just 16 years old, in 1895. He did a thought experiment, where he rode on one light wave and looked at another light wave moving parallel to him.

Classical physics should say that the light wave Einstein was looking at would have a relative speed of zero, but this contradicted Maxwell's equations that showed light always has the same speed: 186,000 miles a second. Another problem with relative speeds is they would show that the laws of electromagnetism change depending on your vantage point, which contradicted classical physics as well (which said the laws of physics were the same for everyone).

This led to Einstein's eventual musings on the theory of special relativity, which he broke down into the everyday example of a person standing beside a moving train, comparing observations with a person inside the train. He imagined the train being at a point in the track equally between two trees. If a bolt of lightning hit both trees at the same time, due to the motion of the train, the person on the train would see the bolt hit one tree before the other tree. But the person beside the track would see simultaneous strikes.

"Einstein concluded that simultaneity is relative; events that are simultaneous for one observer may not be for another,". "This led him to the counterintuitive idea that time flows differently according to the state of motion, and to the conclusion that distance is also relative." Einstein's work led to some startling results, which today still seem counterintuitive at first glance even though his physics is usually introduced at the high school level.

One of the most famous equations in mathematics comes from special relativity. The equation — $E = mc^2$ — means "energy equals mass times the speed of light squared." It shows that energy (E) and mass (m) are interchangeable; they are different forms of the same thing. If mass is somehow totally converted into energy, it also shows how much energy would reside inside that mass: quite a lot. (This equation is one of the demonstrations for why an atomic bomb is so powerful, once its mass is converted to an explosion.)

This equation also shows that mass increases with speed, which effectively puts a speed limit on how fast things can move in the universe. Simply put, the speed of light (c) is the fastest velocity at which an object can travel in a vacuum. As an object moves, its mass also increases. Near the speed of light, the mass is so high that it reaches infinity, and would require infinite energy to move it, thus capping how fast an object can move. The only reason light moves at the speed it does is because photons, the quantum particles that make up light, have a mass of zero.

A special situation in the universe of the small, called "quantum entanglement," is confusing because it seems to involve quantum particles interacting with each other at speeds faster than the speed of light. Specifically, measuring the property of one particle can instantly tell you the property of another particle, no matter how far away they are. Much has been written about this phenomenon, which is still not fully explained in terms of Einstein's conclusions.

Another strange conclusion of Einstein's work comes from the realization that time moves relative to the observer. An object in motion experiences time dilation, meaning that time moves more slowly when one is moving, than when one is standing still. Therefore, a person moving ages more slowly than a person at rest. So yes, when astronaut Scott Kelly spent nearly a year aboard the International Space Station in 2015-16, his twin astronaut brother Mark Kelly aged a little faster than Scott.

This becomes extremely apparent at speeds approaching the speed of light. Imagine a 15-year-old traveling at 99.5 percent the speed of light for five years (from the astronaut's perspective). When the 15-year-old gets back to Earth, according to NASA, he would be only 20 years old. His classmates, however, would be 65 years old.

While this time dilation sounds very theoretical, it does have practical applications as well. If you have a Global Positioning Satellite (GPS) receiver in your car, the receiver attempts to find signals from at least three satellites to coordinate your position. The GPS satellites send out timed radio signals that the receiver listens to, triangulating (or more properly speaking, trilaterating) its position based on the travel time of the signals. The challenge is, the atomic clocks on the GPS are moving and would therefore run faster than atomic clocks on Earth, creating timing issues. So, engineers need to make the clocks on a GPS tick slower.

The clocks in space tick faster, because the GPS satellites are above Earth and experience weaker gravity. So even though the GPS satellites are moving and experience a seven-microsecond slowing every day because of their movement, the result of the weaker gravity causes the clocks to tick about 45 microseconds faster than a ground-based clock. Adding the two together, results in the GPS satellite clock ticking faster than a ground-based clock, by about 38 microseconds daily.

Special Relativity and Quantum Mechanics

As our knowledge of physics has advanced, scientists have run into more counterintuitive situations. One is trying to reconcile general relativity — which describes well what's going on with large objects — with quantum mechanics, which is best used for very small things (such as uranium atom decay). The two fields, which excellently describe their individual fields, are incompatible with one another — which frustrated Einstein and generations of scientists after him.

"Relativity gives nonsensical answers when you try to scale it down to quantum size, eventually descending to infinite values in its description of gravity. Likewise, quantum mechanics runs into serious trouble when you blow it up to cosmic dimensions".

Quantum fields carry a certain amount of energy, even in seemingly empty space, and the amount of energy gets bigger as the fields get bigger. According to Einstein, energy and mass are equivalent (that's the message of $E = mc^2$), so piling up energy is exactly like piling up mass. Go big enough, and the amount of energy in the quantum fields becomes so great that it creates a black hole that causes the universe to fold in on itself.

Lorentz Transformation

Lorentz transformation is the relationship between two different coordinate frames that move at a constant velocity and are relative to each other. The name of the transformation comes from a Dutch physicist Hendrik Lorentz.

We have used the postulates of relativity to examine, in particular examples, how observers in different frames of reference measure different values for lengths and the time intervals. We can gain further insight into how the postulates of relativity change the Newtonian view of time and space by examining the transformation equations that give the space and time coordinates of events in one inertial reference frame in terms of those in another. We first examine how position and time coordinates transform between inertial frames according to the view in Newtonian physics. Then we examine how this has to be changed to agree with the postulates of relativity. Finally, we examine the resulting Lorentz transformation equations and some of their consequences in terms of four-dimensional space-time diagrams, to support the view that the consequences of special relativity result from the properties of time and space itself, rather than electromagnetism.

The Galilean Transformation Equations

An event is specified by its location and time (x, y, z, t) relative to one particular inertial frame of reference S. As an example, (x, y, z, t) could denote the position of a particle at time t, and we could be looking at these positions for many different times to follow the motion of the particle. Suppose a second frame of reference S' moves with velocity v with respect to the first. For simplicity, assume this relative velocity is along the x-axis. The relation between the time and coordinates in the two frames of reference is then,

$$x = x' + vt, y = y', z = z'.$$

Implicit in these equations is the assumption that time measurements made by observers in both S and S' are the same. That is,

$$t = t'.$$

These four equations are known collectively as the Galilean transformation.

We can obtain the Galilean velocity and acceleration transformation equations by differentiating these equations with respect to time. We use u for the velocity of a particle to distinguish it from v, the relative velocity of two reference frames. Note that, for the Galilean transformation, the increment of time used in differentiating to calculate the particle velocity is the same in both frames, $dt = dt'$. Differentiation yields,

$$u_x = u'_x + v, \; u_y = u'_y, u_z = u'_z$$

and,

$$a_x = a'_x, \; a_y = a'_y, \; a_z = a'_z.$$

We denote the velocity of the particle by u rather than v to avoid confusion with the velocity v of one frame of reference with respect to the other. Velocities in each frame differ by the velocity that one frame has as seen from the other frame. Observers in both frames of reference measure the same value of the acceleration. Because the mass is unchanged by the transformation, and distances between points are uncharged, observers in both frames see the same forces $F = ma$ acting between objects and the same form of Newton's second and third laws in all inertial frames. The laws of mechanics are consistent with the first postulate of relativity.

The Lorentz Transformation Equations

The Galilean transformation nevertheless violates Einstein's postulates, because the velocity equations state that a pulse of light moving with speed c along the x-axis would travel at speed $c-v$ in the other inertial frame. Specifically, the spherical pulse has radius $r=ct$ at time t in the unprimed frame, and also has radius $r'=ct'$ at time t' in the primed frame. Expressing these relations in Cartesian coordinates gives:

$$\begin{aligned} x^2+y^2+z^2-c^2t^2 &= 0 \\ x'^2+y'^2+z'^2-c^2t'^2 &= 0 \end{aligned}$$

The left-hand sides of the two expressions can be set equal because both are zero. Because $y=y'$ and $z=z'$, we obtain $x^2-c^2t^2=x'^2-c^2t'^2$.

This cannot be satisfied for nonzero relative velocity v of the two frames if we assume the Galilean transformation results in $t=t'$ with $x=x'+vt'$.

To find the correct set of transformation equations, assume the two coordinate systems S and S' in. First suppose that an event occurs at $(x',0,0,t')$ in S' and at $(x,0,0,t)$ in S, as depicted in the figure.

An event occurs at $(x, 0, 0, t)$ in S and at $(x',0,0,t')$ in S'. The Lorentz transformation equations relate events in the two systems.

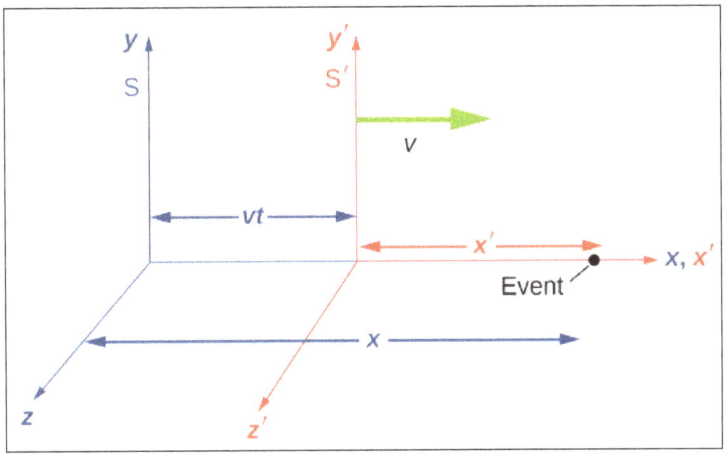

Suppose that at the instant that the origins of the coordinate systems in S and S' coincide, a flash bulb emits a spherically spreading pulse of light starting from the origin. At time t, an observer in S finds the origin of S' to be at $x=vt$. with the help of a friend in S, the S' observer also measures the distance from the event to the origin of S' and finds it to be $x'\sqrt{1-v^2/c^2}$. This follows because we have already shown the postulates of relativity to imply length contraction. Thus the position of the event in S is $x=vt+x'\sqrt{1-v^2/c^2}$,

and,

$$x'=\frac{x-vt}{\sqrt{1-v^2/c^2}}$$

The postulates of relativity imply that the equation relating distance and time of the spherical wave front:

$$x^2 + y^2 + z^2 - c^2t^2 = 0$$

Must apply both in terms of primed and un-primed coordinates, which was shown above to lead to:

$$x^2 - c^2t^2 = x'^2 - c^2t'^2.$$

We combine this with the equation relating x and x' to obtain the relation between t and t':

$$t' = \frac{t - vx/c^2}{\sqrt{1 - v^2/c^2}}.$$

The equations relating the time and position of the events as seen in S are then,

$$\begin{aligned} t &= \frac{t' + vx'/c^2}{\sqrt{1 - v^2/c^2}} \\ x &= \frac{x' + vt'}{\sqrt{1 - v^2/c^2}} \\ y &= y' \\ z &= z'. \end{aligned}$$

This set of equations, relating the position and time in the two inertial frames, is known as the Lorentz transformation. They are named in honour of H.A. Lorentz, who first proposed them. Interestingly, he justified the transformation on what was eventually discovered to be a fallacious hypothesis. The correct theoretical basis is Einstein's special theory of relativity.

The reverse transformation expresses the variables in S in terms of those in S' simply interchanging the primed and un-primed variables and substituting gives:

$$\begin{aligned} t' &= \frac{t - vx/c^2}{\sqrt{1 - v^2/c^2}} \\ x' &= \frac{x - vt}{\sqrt{1 - v^2/c^2}} \\ y' &= y \\ z' &= z. \end{aligned}$$

Space-Time

Relativistic phenomena can be analyzed in terms of events in a four-dimensional space-time. When phenomena such as the twin paradox, time dilation, length contraction, and the dependence of simultaneity on relative motion are viewed in this way, they are seen to be characteristic of the nature of space and time, rather than specific aspects of electromagnetism.

In three-dimensional space, positions are specified by three coordinates on a set of Cartesian axes, and the displacement of one point from another is given by:

$$(x,y,z)=(x_2-x_1,y_2-y_1,z_2-z_1).$$

The distance r between the points is,

$$r\ =(x)^2+(y)^2+(z)^2$$

The distance r is invariant under a rotation of axes. If a new set of Cartesian axes rotated around the origin relative to the original axes are used, each point in space will have new coordinates in terms of the new axes, but the distance r' given by,

$$r'^2=(x')^2+(y')^2+(z')^2.$$

That has the same value that r^2 had. Something similar happens with the Lorentz transformation in space-time.

Define the separation between two events, each given by a set of x, y, z¸ and ct along a four-dimensional Cartesian system of axes in space-time, as:

$$(x,y,z,ct)=(x_2-x_1,y_2-y_1,z_2-z_1,c(t_2-t_1)).$$

Also define the space-time interval s between the two events as,

$$s^2=(x)^2+(y)^2+(z)^2-(ct)^2.$$

If the two events have the same value of *ct* in the frame of reference considered, s would correspond to the distance r between points in space.

The path of a particle through space-time consists of the events (*x*, *y*, *z*¸ *ct*) specifying a location at each time of its motion. The path through space-time is called the world line of the particle. The world line of a particle that remains at rest at the same location is a straight line that is parallel to the time axis. If the particle moves at constant velocity parallel to the *x*-axis, its world line would be a sloped line $x=vt$, corresponding to a simple displacement vs. time graph. If the particle accelerates, its world line is curved. The increment of s along the world line of the particle is given in differential form as:

$$ds^2=(dx)^2+(dy)^2+(dz)^2-c^2(dt)^2.$$

Just as the distance r is invariant under rotation of the space axes, the space-time interval:

$$s^2=(x)^2+(y)^2+(z)^2-(ct)^2.$$

is invariant under the Lorentz transformation. This follows from the postulates of relativity, and can be seen also by substitution of the previous Lorentz transformation equations into the expression for the space-time interval:

$$s^2=(x)^2+(y)^2+(z)^2-(ct)^2$$

$$= \left(\frac{x' + vt'}{\sqrt{1 - v^2/c^2}} \right)^2 + (y')^2 + (z')^2 - \left(c \frac{t' + \frac{vx'}{c^2}}{\sqrt{1 - v^2/c^2}} \right)^2$$

$$= (x')^2 + (y')^2 + (z')^2 - (ct')^2$$

$$= s'^2.$$

In addition, the Lorentz transformation changes the coordinates of an event in time and space similarly to how a three-dimensional rotation changes old coordinates into new coordinates:

Lorentz transformation (x, t coordinates)	Axis – rotation around z-axis (x, y coordinates)
$x' = (\gamma)x + (\beta\gamma)ct$	$x' = (\cos\theta)x + (\sin\theta)y$
$ct' = (-\beta\gamma)x + (\gamma)ct$	$y'(-\sin\theta)x + (\cos\theta)y$

Where $\gamma = \frac{1}{\sqrt{1-\beta^2}}; \beta = v/c.$

Lorentz transformations can be regarded as generalizations of spatial rotations to space-time. However, there are some differences between a three-dimensional axis rotation and a Lorentz transformation involving the time axis, because of differences in how the metric, or rule for measuring the displacements r and s, differ. Although r is invariant under spatial rotations and s is invariant also under Lorentz transformation, the Lorentz transformation involving the time axis does not preserve some features, such as the axes remaining perpendicular or the length scale along each axis remaining the same.

The quantity s^2 can have either sign, depending on the coordinates of the space-time events involved. For pairs of events that give it a negative sign, it is useful to define τ^2 as $-s^2$. The significance of τ as just defined follows by noting that in a frame of reference where the two events occur at the same location, we have $x = y = z = 0$ and therefore (from the equation for $s^2 = -\tau^2$:

$$\tau^2 = -s^2 = (t)^2.$$

Therefore τ is the time interval t in the frame of reference where both events occur at the same location. It is the same interval of proper time. It also follows from the relation between s and that τ that because s is Lorentz invariant, the proper time is also Lorentz invariant. All observers in all inertial frames agree on the proper time intervals between the same two events.

The Light Cone

We can deal with the difficulty of visualizing and sketching graphs in four dimensions by imagining the three spatial coordinates to be represented collectively by a horizontal axis, and the vertical axis to be the ct-axis. Starting with a particular event in space-time as the origin of the space-time graph shown, the world line of a particle that remains at rest at the initial location of the event at the origin then is the time axis. Any plane through the time axis parallel to the spatial axes contains all the events that are simultaneous with each other and with the intersection of the plane and the time axis, as seen in the rest frame of the event at the origin.

It is useful to picture a light cone on the graph, formed by the world lines of all light beams passing through the origin event A, as shown in. The light cone, according to the postulates of relativity, has sides at an angle of 45 if the time axis is measured in units of ct, and, according to the postulates of relativity, the light cone remains the same in all inertial frames. Because the event A is arbitrary, every point in the space-time diagram has a light cone associated with it. The light cone consists of all the world lines followed by light from the event A at the vertex of the cone.

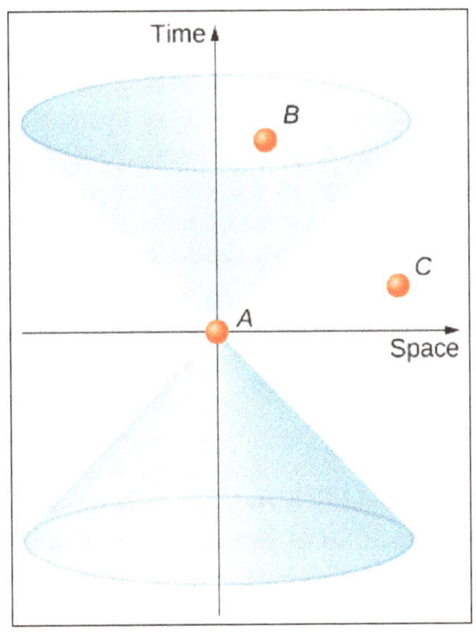

Consider now the world line of a particle through space-time. Any world line outside of the cone, such as one passing from A through C, would involve speeds greater than c, and would therefore not be possible. Events such as C that lie outside the light cone are said to have a space-like separation from event A. They are characterized by:

$$s_{AC}^2 = (x_A - x_B)^2 + (x_A - x_B)^2 + (x_A - x_B)^2 - (ct)^2 > 0.$$

An event like B that lies in the upper cone is reachable without exceeding the speed of light in vacuum, and is characterized by:

$$s_{AB}^2 = (x_A - x_B)^2 + (x_A - x_B)^2 + (x_A - x_B)^2 - (ct)^2 < 0.$$

The event is said to have a time-like separation from A. Time-like events that fall into the upper half of the light cone occur at greater values of t than the time of the event A at the vertex and are in the future relative to A. Events that have time-like separation from A and fall in the lower half of the light cone are in the past, and can affect the event at the origin. The region outside the light cone is labeled as neither past nor future, but rather as "elsewhere".

For any event that has a space-like separation from the event at the origin, it is possible to choose a time axis that will make the two events occur at the same time, so that the two events are simultaneous in some frame of reference. Therefore, which of the events with space-like separation comes before the other in time also depends on the frame of reference of the observer. Since space-like separations can be traversed only by exceeding the speed of light; this violation of which event can cause the other provides another argument for why particles cannot travel faster than the speed of light, as well as potential material for science fiction about time travel. Similarly for any event with time-like separation from the event at the origin, a frame of reference can be found that will make the events occur at the same location. Because the relations,

$$s_{AC}^2 = (x_A - x_B)^2 + (x_A - x_B)^2 + (x_A - x_B)^2 - (ct)^2 > 0$$

and,

$$s_{AB}^2 = (x_A - x_B)^2 + (x_A - x_B)^2 + (x_A - x_B)^2 - (ct)^2 < 0.$$

are Lorentz invariant, whether two events are time-like and can be made to occur at the same place or space-like and can be made to occur at the same time is the same for all observers. All observers in different inertial frames of reference agree on whether two events have a time-like or space-like separation.

The Twin Paradox seen in Space-Time

The twin paradox involves an astronaut twin traveling at near light speed to a distant star system, and returning to Earth. Because of time dilation, the space twin is predicted to age much less than the earthbound twin. This seems paradoxical because we might have expected at first glance for the relative motion to be symmetrical and naively thought it possible to also argue that the earthbound twin should age less.

To analyze this in terms of a space-time diagram, assume that the origin of the axes used is fixed in Earth. The world line of the earthbound twin is then along the time axis. The world line of the astronaut twin, who travels to the distant star and then returns, must deviate from a straight line path in order to allow a return trip. As seen in, the circumstances of the two twins are not at all symmetrical. Their paths in space-time are of manifestly different length. Specifically, the world line of the earthbound twin has length $2ct$, which then gives the proper time that elapses for the earthbound twin as $2t$. The distance to the distant star system is $x = vt$. the proper time that elapses for the space twin is 2τ where,

$$c^2\tau^2 = -s^2 = (ct)^2 - (x)^2.$$

This is considerably shorter than the proper time for the earthbound twin by the ratio:

$$\frac{c\tau}{ct} = \sqrt{\frac{(ct)^2 - (x)^2}{(ct)^2}} = \sqrt{\frac{(ct)^2 - (vt)^2}{(ct)^2}}$$
$$= \sqrt{1 - \frac{v^2}{c^2}} = \frac{1}{\gamma}.$$

Consistent with the time dilation formula. The twin paradox is therefore seen to be no paradox at all. The situation of the two twins is not symmetrical in the space-time diagram. The only surprise is perhaps that the seemingly longer path on the space-time diagram corresponds to the smaller proper time interval, because of how τ and s depend on x and t. The space twin and the earthbound twin, in the twin paradox example, follow world lines of different length through space-time.

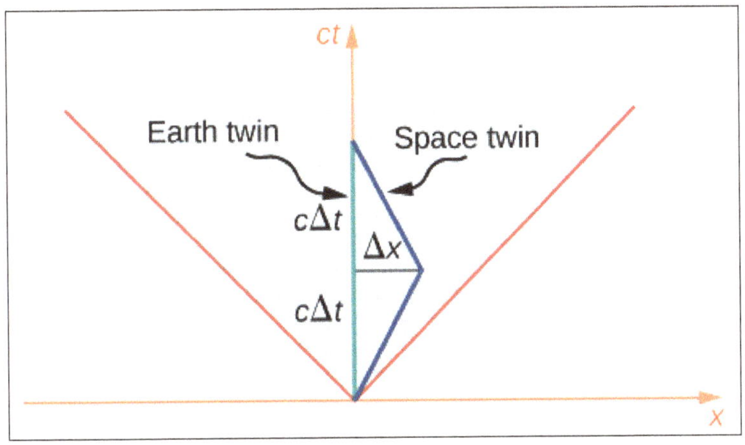

Lorentz Transformations in Space-Time

We have already noted how the Lorentz transformation leaves:

$$s^2 = (x)^2 + (y)^2 + (z)^2 - (ct)^2$$

Un-changed and corresponds to a rotation of axes in the four-dimensional space-time. If the S and S' frames are in relative motion along their shared x-direction the space and time axes of S' are rotated by an angle α as seen from S, in the way shown in shown in, where:

$$\tan\alpha = \frac{v}{c} = \beta.$$

This differs from a rotation in the usual three-dimension sense, insofar as the two space-time axes rotate toward each other symmetrically in a scissors-like way, as shown. The rotation of the time and space axes is both through the same angle. The mesh of dashed lines parallel to the two axes show how coordinates of an event would be read along the primed axes. This would be done by following a line parallel to the x' and one parallel to the t'-axis, as shown by the dashed lines. The length scale of both axes is changed by:

$$ct' = ct\sqrt{\frac{1+\beta^2}{1-\beta^2}}; x' = x\sqrt{\frac{1+\beta^2}{1-\beta^2}}.$$

The line labelled "v = c" at 45 to the x-axis corresponds to the edge of the light cone, and is unaffected by the Lorentz transformation, in accordance with the second postulate of relativity. The "v = c" line, and the light cone it represents, are the same for both the S and S' frame of reference.

The Lorentz transformation results in new space and time axes rotated in a scissors-like way with respect to the original axes.

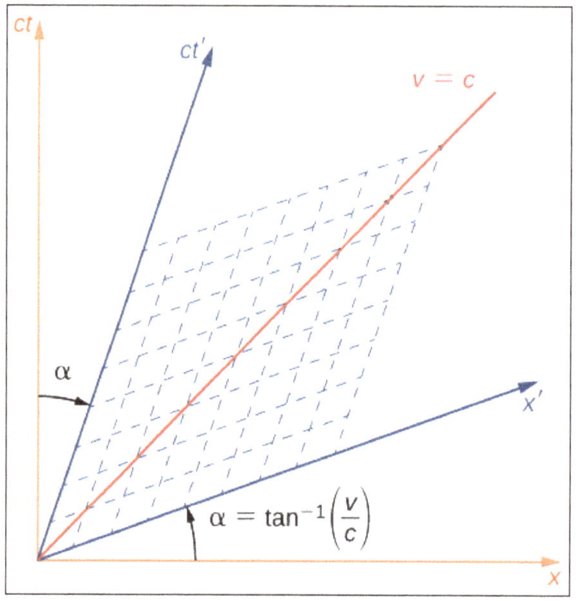

Simultaneity

Simultaneity of events at separated locations depends on the frame of reference used to describe them, as given by the scissors-like "rotation" to new time and space coordinates as described. If two events have the same t values in the un-primed frame of reference, they need not have the same values measured along the ct'-axis, and would then not be simultaneous in the primed frame.

As a specific example, consider the near-light-speed train in which flash lamps at the two ends of the car have flashed simultaneously in the frame of reference of an observer on the ground. The space-time graph is shown. The flashes of the two lamps are represented by the dots labelled "Left flash lamp" and "Right flash lamp" that lie on the light cone in the past. The world line of both pulses travel along the edge of the light cone to arrive at the observer on the ground simultaneously. Their arrival is the event at the origin. They therefore had to be emitted simultaneously in the un-primed frame, as represented by the point labelled as t(both). But time is measured along the ct'-axis, in the frame of reference of the observer seated in the middle of the train car. So in her frame of reference, the emission event of the bulbs labelled as t' (left) and t' (right), were not simultaneous. The train example revisited. The flashes occur at the same time t(both) along the time axis of the ground observer, but at different times, along the t' time axis of the passenger.

In terms of the space-time diagram, the two observers are merely using different time axes for the same events because they are in different inertial frames, and the conclusions of both observers are equally valid. As the analysis in terms of the space-time diagrams further suggests, the property of how simultaneity of events depends on the frame of reference results from the properties of space and time itself, rather than from anything specifically about electromagnetism.

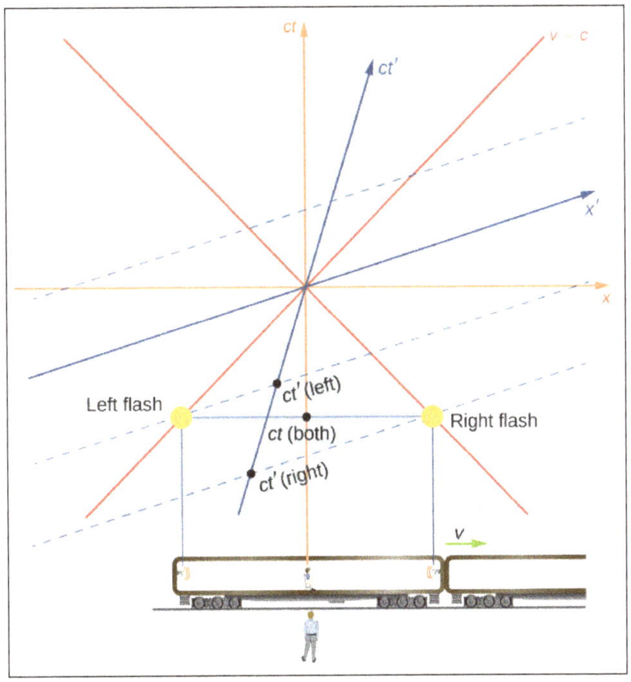

Problems

Describe the following physical occurrences as events, that is, in the form (x, y, z, t): (a) A postman rings a doorbell of a house precisely at noon. (b) At the same time as the doorbell is rung, a slice of bread pops out of a toaster that is located 10 m from the door in the east direction from the door. (c) Ten seconds later, an airplane arrives at the airport, which is 10 km from the door in the east direction and 2 km to the south.

Describe what happens to the angle $\alpha = \tan(v/c)$, and therefore to the transformed axes in, as the relative velocity v of the S and S' frames of reference approaches c.

The angle α approaches 45 and the t'- and x'-axes rotate toward the edge of the light cone.

Describe the shape of the world line on a space-time diagram of (a) an object that remains at rest at a specific position along the x-axis; (b) an object that moves at constant velocity u in the x-direction; (c) an object that begins at rest and accelerates at a constant rate of in the positive x-direction.

A man standing still at a train station watches two boys throwing a baseball in a moving train. Suppose the train is moving east with a constant speed of 20 m/s and one of the boys throws the ball with a speed of 5 m/s with respect to himself toward the other boy, who is 5 m west from him. What is the velocity of the ball as observed by the man on the station?

15 m/s east

When observed from the sun at a particular instant, Earth and Mars appear to move in opposite directions with speeds 108,000 km/h and 86,871 km/h, respectively. What is the speed of Mars at this instant when observed from Earth?

A man is running on a straight road perpendicular to a train track and away from the track at a speed of 12 m/s. The train is moving with a speed of 30 m/s with respect to the track. What is the speed of the man with respect to a passenger sitting at rest in the train?

32 m/s

A man is running on a straight road that makes 30 with the train track. The man is running in the direction on the road that is away from the track at a speed of 12 m/s. The train is moving with a speed of 30 m/s with respect to the track. What is the speed of the man with respect to a passenger sitting at rest in the train?

In a frame at rest with respect to the billiard table, a billiard ball of mass m moving with speed v strikes another billiard ball of mass m at rest. The first ball comes to rest after the collision while the second ball takes off with speed v in the original direction of the motion of the first ball. This shows that momentum is conserved in this frame. (a) Now, describe the same collision from the perspective of a frame that is moving with speed v in the direction of the motion of the first ball. (b) Is the momentum conserved in this frame?

a. The second ball approaches with velocity −v and comes to rest while the other ball continues with velocity −v; b. This conserves momentum.

In a frame at rest with respect to the billiard table, two billiard balls of same mass m are moving toward each other with the same speed v. After the collision, the two balls come to rest. (a) Show that momentum is conserved in this frame. (b) Now, describe the same collision from the perspective of a frame that is moving with speed v in the direction of the motion of the first ball. (c) Is the momentum conserved in this frame?

In a frame S, two events are observed: event 1: a pion is created at rest at the origin and event 2: the pion disintegrates after time τ. Another observer in a frame S' is moving in the positive direction along the positive x-axis with a constant speed v and observes the same two events in his frame. The origins of the two frames coincide at $t = t' = 0$. (a) Find the positions and timings of these two events in the frame S' (a) according to the Galilean transformation, and (b) according to the Lorentz transformation.

a. $\begin{matrix} t_1' = 0; & x_1' = 0; \\ t_2' = \tau; & x_2' = 0 \end{matrix}$; b. $\begin{matrix} t_1' = 0; \quad x_1' = 0; \\ t_2' = \dfrac{\tau}{\sqrt{1 - v^2/c^2}}; x_2' = \dfrac{-v\tau}{\sqrt{1 - v^2/c^2}} \end{matrix}$

General Relativity

The general theory of relativity describes the force of gravity. Einstein wasn't the first to come up with such a theory — back in 1686 Isaac Newton formulated his famous inverse square law of gravitation. Newton's law works perfectly well on small-ish scales: we can use it to calculate how fast an object dropped off a tall building will hurtle to the ground and even to send people to the Moon. But when distances and speeds are very large, or very massive objects are involved, Newton's law becomes inaccurate. It's a good place to start though, as it's easier to describe than Einstein's theory.

Suppose you have two objects, say the Sun and the Earth, with masses m_1 and m_2 respectively.

Write r for the distance between the two objects. Then Newton's law says that the gravitational force F between them is:

$$F = G\frac{m_1 m_2}{r^2},$$

Where, G is a fixed number, known as Newton's constant.

The formula makes intuitive sense: it tells us that gravity gets weaker over long distances (the larger r the smaller F) and that the gravitational force is stronger between more massive objects (the larger either of m_1 and m_2 the larger F).

Different Force Same Formula

There is another formula which looks very similar, but describes a different force. In 1785 the French physicist Charles-Augustin de Coulomb came up with an equation to capture the electrostatic force F that acts between two charged particles with charges Q_1 and Q_2:

$$F = \frac{1}{4\pi\epsilon_0}\frac{Q_1 Q_2}{r^2}.$$

Here r stands for the distance between the two particles and ϵ_0 is a constant which determines the strength of electromagnetism. (It has the fancy name permittivity of free space.)

The Problem with Newton

Newton's and Coulomb's formulas are nice and neat, but there is a problem. Going back to Newton's law, suppose you took the Earth and the Sun and very quickly moved them further apart. This would make the force acting between them weaker, but, according to the formula, the weakening of the force would happen straight away, the instant you move the two bodies apart. The same goes for Coulomb's law: moving the charged particles apart very quickly would result in an immediate weakening of the electrostatic force between them.

But this can't be true. Einstein's special theory of relativity, proposed ten years before the general theory in 1905, says that nothing in the Universe can travel faster than light — not even the "signal" that communicates that two objects have moved apart and the force should become weaker.

Why we Need Fields

This is one reason why the classical idea of a force needs replacing in modern physics. Instead, we need to think in terms of something — new objects — that transmit the force between one object and another. This was the great contribution of the British scientist Michael Faraday to theoretical physics. Faraday realized that spread throughout the Universe there are objects we today call fields, which are involved in transmitting a force. Examples are the electric and magnetic fields you are probably familiar with from school.

A charged particle gives rise to an electric field, which is "felt" by another particle (which has its own electric field). One particle will move in response to the other's electric field — that's what we call a force. When one particle is quickly moved away from the other, then this causes ripples in the first particle's electric field. The ripples travel through space, at the speed of light, and eventually affect the other particle. In fact, the particle that is moved also generates a magnetic field and emits electromagnetic radiation. The end result is a complex interaction of rippling fields — but the point is that the force is really one particle being affected by ripples propagating through the field of the other.

It took scientists a long time to fully develop this field picture of electromagnetism. The main credit goes to the Scottish scientist James Clerk Maxwell, who not only realized that the electric and magnetic forces were two aspects of a unified force of electromagnetism, but also replaced Coulomb's simple law of electrostatics with four equations that describe how electric and magnetic fields respond to moving charged particles. Maxwell's four formulas are some of the most amazing equations in physics because they capture all there is to know about electricity and magnetism.

Albert Einstein in 1921.

Gravity and Spacetime

So what about gravity? Just as with electromagnetism there needs to be a field giving rise to what we perceive as the gravitational force acting between two bodies. Einstein's great insight was that this field is made of something we already know about: space and time. Imagine a heavy body, like the Sun, sitting in space. Einstein realized that space isn't just a passive by-stander, but responds to the heavy object by bending. Another body, like the Earth, moving into the dent created by the heavier object will be diverted by that dent. Rather than carrying on moving along a straight line, it will start orbiting the heavier object. Or, if it is sufficiently slow, will crash into it.

Einstein's theory is that space and time can warp into each other — they are inextricable linked and time, too, can be distorted by massive objects. This is why we talk, not just about the curvature of space, but about the curvature of spacetime.

The Equation

The general theory of relativity is captured by a deceptively simple-looking equation:

$$R_{\mu\nu} - \frac{1}{2} R g_{\mu\nu} = \frac{8\pi G}{c^4} T_{\mu\nu}.$$

Essentially the equation tells us how a given amount of mass and energy warps spacetime. The left-hand side of the equation,

$$R_{\mu\nu} - \frac{1}{2} R g_{\mu\nu},$$

Describes the curvature of spacetime whose effect we perceive as the gravitational force. It's the analogue of the term F on the left-hand side of Newton's equation.

Massive objects bend spacetime.

The term $T_{\mu\nu}$ on right-hand side of the equation describes everything there is to know about the way mass, energy, momentum and pressure are distributed throughout the Universe. It is what became of the term $m_1 m_2$ in Newton's equation, but it is much more complicated. All of these things are needed to figure out how space and time bend. $T_{\mu\nu}$ goes by the technical term energy-momentum tensor. The constant G that appears on the right-hand side of the equation is again Newton's constant and c is the speed of light.

What about the Greek letters μ and ν that appear as subscripts? To understand what they mean, first notice that spacetime has four dimensions. There are three dimensions of space (corresponding to the three directions left-right, up-down and forwards-backwards of space) and one dimension of time (which only has one direction). If you want to understand how a moving bit of mass affects spacetime, you need to understand how it affects each of those four dimensions and their various combinations.

(As an analogy, think of the way you'd describe an object moving at constant speed along a straight line in Newton's classical physics. You need two pieces of information: the direction and the speed of the motion. The direction is given by three numbers, each telling you by how much the object moves in each of the three directions of space. Therefore, the motion is described by a total of four numbers, three relating to space and one giving the speed. Since speed is distance covered per unit time, we need three bits of information relating to space and one to time, in order to describe the motion).

Not Just One Equation

In Einstein's equation the Greek letters μ and ν are labels, which can each take on the values 0, 1, 2 or 3. So really, the equation above conceals a whole collection of equations corresponding to the possible combinations of values μ and ν can take:

$$R_{00} - \frac{1}{2} R g_{00} = \frac{8\pi G}{c^4} T_{00}$$

$$R_{01} - \frac{1}{2} R g_{01} = \frac{8\pi G}{c^4} T_{01}$$

$$R_{11} - \frac{1}{2} R g_{11} = \frac{8\pi G}{c^4} T_{11}$$

And so on.

The value of 0 corresponds to time and the values 1, 2 and 3 to the three dimensions of space.

The equation:

$$R_{01} - \frac{1}{2} R g_{01} = \frac{8\pi G}{c^4} T_{01}$$

Therefore relates to time and the 1-direction of space. The T term on the right-hand side describes the momentum (speed and mass) of matter moving in the 1-direction of space. The motion causes time and the 1-direction of space to mix and warp into each other — that effect is described by the left-hand side of the equation. (The analogue goes for an equation with $\mu = 0$ and ν equal to 2 or 3.)

If the equation only has 1s, 2s or 3s, for example,

$$R_{11} - \frac{1}{2} R g_{11} = \frac{8\pi G}{c^4} T_{11},$$

Then it relates only to space. The T term on the right-hand side now measures the pressure that matter causes in the corresponding direction of space. The left-hand side tells you how that matter causes space in that direction to stretch.

If μ and ν both take the value 0, then the equation

$$R_{00} - \frac{1}{2} R g_{00} = \frac{8\pi G}{c^4} T_{00}$$

Only relates to time. The term T_{00} now stands for energy, which causes time to speed up or slow down. The left-hand side of the equation describes that change in the flow of time.

Since each of μ and ν can take on four values, this gives a total of 4 x 4 =16 equations. However, it turns out that the equation with $\mu = i$ and $\nu = j$ (for i and j each equal to one of 0, 1, 2 or 3) is the same as the equation with $\mu = j$ and $\nu = i$. this reduces the total number of equations to ten.

An artist's impression of a black hole.

In theory Einstein's equations allow you to work out exactly how massive objects, such as planets, stars, galaxies, or even black holes affect the spacetime they sit in. In practice though, things aren't quite as straight-forward. Einstein's equations are incredibly difficult to solve – supercomputers are needed to find solutions and coming up with new solutions is an active field of theoretical physics. One of the big current challenges is to figure out what happens to space-time when two very heavy objects, like black holes, collide.

How do we know that Einstein's theory is correct? In the hundred years since its publication, the theory has passed every test it has been subjected to. Despite its slightly esoteric nature, it's crucial in things most of us rely on every day, such as the GPS features in our smartphones and the Satnav devices in our cars. The theory does open up some new questions, which is why some physicists think it needs to be modified. But whether or not this really turns out to be necessary, there's no doubt that general relativity is one of the most amazing achievements in the history of science.

Einstein Field Equation

The Einstein Field Equation (EFE) is also known as Einstein's equation. There are ten nonlinear partial differential equations of Einstein field extracted from Albert Einstein's General Theory of Relativity. The EFE describes the basic interaction of gravitation. The equations were first published in 1915 by Albert Einstein as a tensor equation.

Following is the Einstein Field Equation:

$$G_{\mu\upsilon} + g_{\mu\upsilon}\Lambda = \frac{8\pi G}{c^4} T_{\mu\upsilon}$$

Where,

- $G_{\mu\upsilon}$ is the Einstein tensor which is given as $R_{\mu\upsilon}$ -½ $Rg\mu\upsilon$.

- $R_{\mu\upsilon}$ is the Ricci curvature tensor.
- R is the scalar curvature.
- $g_{\mu\upsilon}$ is the metric tensor.
- Λ is a cosmological constant.
- G is Newton's gravitational constant.
- c is the speed of light.
- $T_{\mu\upsilon}$ is the stress-energy tensor.

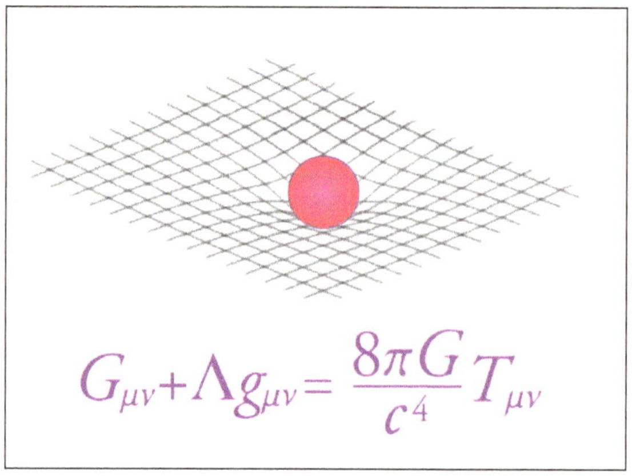

Einstein Field Equations Derivation

Following is the derivation of Einstein Field Equations. Einstein wanted to explain that measure of curvature = source of gravity.

The source of gravity is the stress-energy tensor. The stress-energy tensor is given as:

$$T^{\alpha\beta}=\begin{bmatrix}\rho & 0 & 0 & 0\\ 0 & P & 0 & 0\\ 0 & 0 & P & 0\\ 0 & 0 & 0 & P\end{bmatrix}\rightarrow\begin{bmatrix}\rho & 0 & 0 & 0\\ 0 & 0 & 0 & 0\\ 0 & 0 & 0 & 0\\ 0 & 0 & 0 & 0\end{bmatrix}$$

In the above matrix we see that the P is tending to zero because, for Newton's gravity, the mass density is the source of gravity.

The equation of motion is given as:

$$\frac{du^i}{d\tau}+\Gamma^i_{\upsilon\alpha}u^\upsilon u^\alpha=0$$

$$\frac{du^i}{d\tau}+\Gamma^{00}_i=0$$

$$\frac{du^i}{d\tau}+\frac{1}{2}\frac{\partial g_{00}}{\partial x^i}=0$$

$$\frac{du^i}{d\tau}+\frac{\partial\phi}{\partial x^i}=0$$
$$g_{00}=-(1+2\phi)$$

But we know that $\nabla^2\phi=4\pi G\rho$

Therefore,

$$R^{\mu\upsilon}=-8\pi GT^{\mu\upsilon}$$

Where $-8\pi GT^{\mu\upsilon}$ is constant.

What is Einstein Tensor?

Einstein tensor is also known as trace-reversed Ricci tensor. In Einstein Field Equation, it is used for describing spacetime curvature such that it is in alignment with the conservation of energy and momentum. It is defined as:

$$G = R\ ½ - gR$$

Where,

- R is the Ricci tensor.
- g is the metric tensor.
- R is the scalar curvature.

What is Stress-Energy Tensor?

Stress-energy tensor is defined as the tensor $T^{\alpha\beta}$ is a symmetrical tensor which is used for describing the energy and momentum density of the gravitational field. It is given as:

$$T^{\alpha\beta} = T^{\beta\alpha}$$

Equivalence Principle

In the general theory of relativity developed by Albert Einstein, the equivalence principle explains about "equivalence of inertial mass and gravitational mass". He observed that the force due to gravity (gravitational force) experienced by a person standing on the massive object (for example earth) is equivalent to the pseudo-force experienced by the observer in the non-inertial frame of reference (accelerated).

In the theory of general relativity, the equivalence principle is the equivalence of gravitational and inertial mass, and Albert Einstein's observation that the gravitational "force" as experienced locally while standing on a massive body (such as the Earth) is the same as the pseudo-force experienced by an observer in a non-inertial (accelerated) frame of reference.

Albert Einstein realized that there is some complexity between Newton's theory of gravity and his own theory of special relativity. He looks into the universe in a completely different way!

As we know, during the year 1905 Albert Einstein set a milestone in physics with his new framework on laws of physics, which is his theory of special relativity. Nonetheless, one facet of physics seemed to be incomplete; the gravitational force theory developed by Newton. The special theory of relativity enfolded space and time excluding gravity. Years later, Albert Einstein managed to unify gravity with his relativistic ideas on space and time. As a result, another revolutionary theory emerged- "The general theory of relativity".

The very first step towards a general theory of relativity was the realization that, even within a gravitational field, there exists a reference frame in which gravity is almost zero/absent. Thus, at least to a certain approximation, one can say gravity-free laws of special relativity govern the laws of physics, if and only if the observations are confined within a small region of space and time. This derived from what he formulated as equivalence principle which, in turn, is inspired by the consequence of a free fall.

What is Effective Mass?

Mass is a property of any physical object, It is a measure of resistance to the acceleration in the presence of external force. Effective mass is the terminology used in the general theory of relativity which talks about gravitational mass and inertial mass.

Gravity: Tendency of massive object to attract each other. Gravitational force between two bodies is given by equation:

$$F = G\frac{Mm_{gravity}}{r^2}$$

Where,

- G is the universal gravitational constant.
- M is the mass of one body.
- $m_{gravity}$ is the mass of the second body.
- r is the distance between them.

Inertia: Resistance offered by the body against the change in its current state of motion. Force involved here is given by equation:

$$F = m_{inertia} \times a$$

Where,

- $m_{inertia}$ is the inertial mass of the object.
- a is the acceleration of the object.

Principle of equivalence comes up with an idea, saying equation $F = G\frac{Mm_{gravity}}{r^2}$ and $F = m_{inertia} \times a$ are the same. That is:

$$F = G\frac{Mm_{gravity}}{r^2} = m_{inertia} \times a \Rightarrow G\frac{Mm_{gravity}}{r^2} = m_{inertia} \times a$$

On rearranging the above equation we get,

$$\Rightarrow a = \left[\frac{m_{gravity}}{m_{inertia}}\right] \times G\frac{M}{r^2}$$

Hence, proves their equivalence mathematically.

Example of Equivalence Principle

Imagine you are standing in an elevator or, precisely, is something inside which looks very similar to an elevator cabin. You are totally isolated from the outer world. Under these circumstances, if you take an object and drop it, it falls down and reaches the floor/base. This is the most expected way you expect, considering your experience here on earth. The situation is depicted in the diagram. Will the effect be the same if the elevator is subjected under the gravitational field, similar to that of the earth?

Theoretically, you may be situated in deep space, far away from the gravitational influence of any significant mass concentration. The cabin you are standing in can be inside a rocket which is accelerating at 9.8 meters per square second. The situation is illustrated clearly in the diagram.

In such a scenario, you drop off an object, The object would experience an acceleration of 9.8 meters per square second from cabin floor- exactly the same acceleration as that an object experience during free fall here, right on earth. Now as you, as an observer standing inside a cabin floor can hardly differentiate two situations. From your frame of reference situations will be indistinguishable: are they falling under the gravity of a massive object like earth? Or is it the cabin floor accelerating them?

Thus, Einstein formulates his Principle of equivalence stating that "No experiment can be performed that could distinguish between a uniform gravitational field and an equivalent uniform acceleration."

Covariance Principle

The term covariance implies a formalism in which the laws of physics maintain the same form under a specified set of transformations.

Classical Mechanic	Physical laws covariant under Galilean coordinate transformations.
Special Relativity	Physical laws covariant under Lorentz coordinate transformations.
General Relativity	Form of physical laws invariant under any arbitrary differentiable coordinate transformation (general covariance).

Special Relativity

One of the two fundamental postulates of special relativity is the principle of relativity, which stipules that the laws of physics are the same in any inertial frame of reference, which is equivalent to say that the laws are covariant by Lorentz transformations.

General Relativity

As we know, in presence of a gravitational field, the Equivalence Principle allows to formulate all - but gravitational - physical laws, in small enough free-falling areas. To get a macroscopic significant description of these laws, one has therefore to find an operation to link these entire different local inertial referential. That will be to introduce a unique non inertial coordinate system, and that is precisely during this coordinate transformation operation that the effects of gravitational field will re-appear.

Some kind of 'inertial' forces in case of an accelerated referential or gravitional effect in case of the presence of gravitational field in a rest frame (Equivalence Principle) was now acting in the new arbitrary referential.

$$f^{\beta} = -m\Gamma^{\beta}_{\mu\nu}\frac{dx^{\mu}}{d\tau}\frac{dx^{\nu}}{d\tau}$$

The idea of Einstein was to set the Principle of General Covariance which extends the principle of relativity to say that the form of the laws of physics should be the same in all - inertial and accelerating - frames.

Therefore, to write a valid physical equation in general relativity, we have to write a Tensor equation (which preserves its form under general coordinate transformations) which is true in special relativity, that's all.

By example, the fundamental Newton's first law of motion is stated as follow in special relativity:

$$F\mu = \frac{Dp^{\mu}}{d\tau}$$

If we postulate with Einstein the general covariance of the physical laws in any arbitrary referential, then we get the correct physic laws by replacing the usual derivatives by the covariant derivatives, end of the story.

The way of writing physics in general relativity or curved spacetime consists then to operate the formal substitution, known as the 'comma goes to semi-colon' rule, as we saw when looking at Covariant Differentiation, commas and semi-colons can be used as shorthand notation for partial and covariant derivatives.

$$, \mapsto ;$$

To be more precise, we will satisfy the Principle of General Covariance by the following rule:

Take the equations of Special Relativity, replace $\eta_{\alpha\beta}$ by $\eta_{\alpha\beta}$ and all derivatives by covariant derivatives.

- This last rule will be applicable for everything except gravity, for which there is no Special Relativity theory.
- To practically use this rule, we have to explicitly calculate the covariant derivative and therefore the Metric tensor has to be known (as the connection coefficient is a function of the met-ric and its derivatives.

Tests of General Relativity

The galactic centre with its supermassive Black Hole Sgr A* and its extreme gravitational potential offers a unique possibility to study effects of general relativity. We use the stars in close orbits around the black hole, especially S2, to find deviations from Newtonian gravity and test General Relativity in different ways.

Measurement of Gravitational Redshift in the Galactic Centre

Sagittarius A* (SgrA*), the massive black hole in the centre of our galaxy is at a distance of 26'000 light years from Earth the closest of its kind and with an apparent Schwarzschild radius of 53μm the largest in the sky. It is surrounded by a cluster of high velocity stars called the S-stars whose trajectories are governed by the gravitational field of the black hole.

Gravitational redshift is one of the three classical tests of General Relativity. Einstein's theory predicts that due to gravitational time dilation a light beam gets stretched to longer wavelengths by a gravitational field. The change in the wavelength of light from S2/S-02 is inconsistent with Newtonian predictions and in excellent agreement with Einstein's theory of general relativity. On a technological level, the success of this measurement with GRAVITY/VLT opens the door to an entirely new type of laboratory to probe and test the General Theory of Relativity: the Galactic Centre.

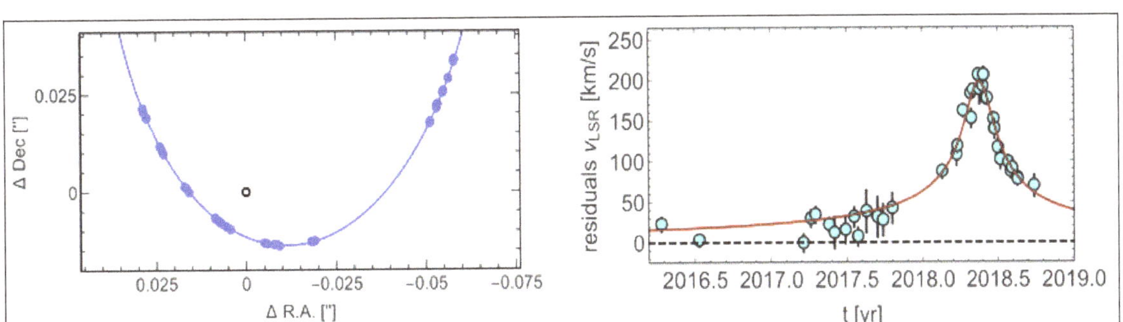

Astrometric and spectroscopic measurements. Left: Position of the star (blue dots) on its trajectory around the black hole (empty circle). The depicted motion is counter clockwise from early 2017 to late 2018. Right: Velocity difference between the Newtonian and relativistic models (red curve) and residuals (circles).

Einstein Equivalence Principle

One of the cornerstones of general relativity is the Einstein Equivalence Principle. It consists of three parts: the weak equivalence principle, the local Lorentz invariance and the local position invariance.

The star S2 experiences very strong changes in gravitational potential on its eccentric orbit around the supermassive black hole Sgr A*. In the three years leading up to the pericenter passage of S2, the gravitational potential experienced by the star changes by four orders of magnitude, which makes it a unique probe and allows testing the LPI. This is possible by observing the redshift from two different atomic lines, Hydrogen and Helium. During the pericenter passage we did not detect a different behaviour of the two absorption lines, which allows us to put a limit on violations of the LPI to below 5%. While current tests on Earth are much more restrictive on that limit, our test in the galactic centre laboratory, using stellar spectra and a supermassive black hole, charts a completely untested potential regime.

In an outlook we also show that with the next generation class of telescopes, such as the ELT, much more restrictive tests of the LPI and also the weak equivalence principle will be possible.

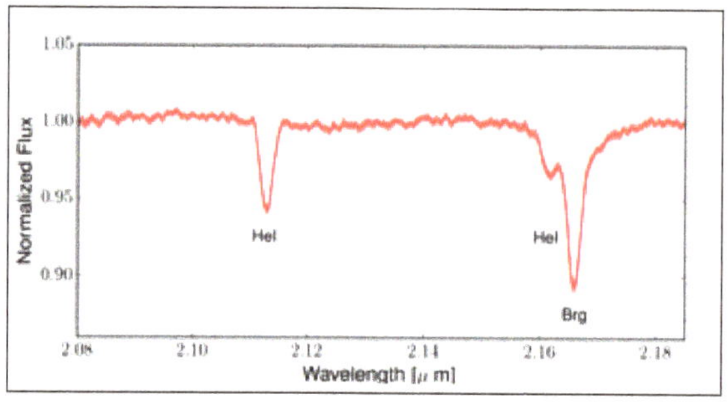

High signal-to-noise spectrum of S2, which shows strong atomic line from Hydrogen (Brg) and Helium.

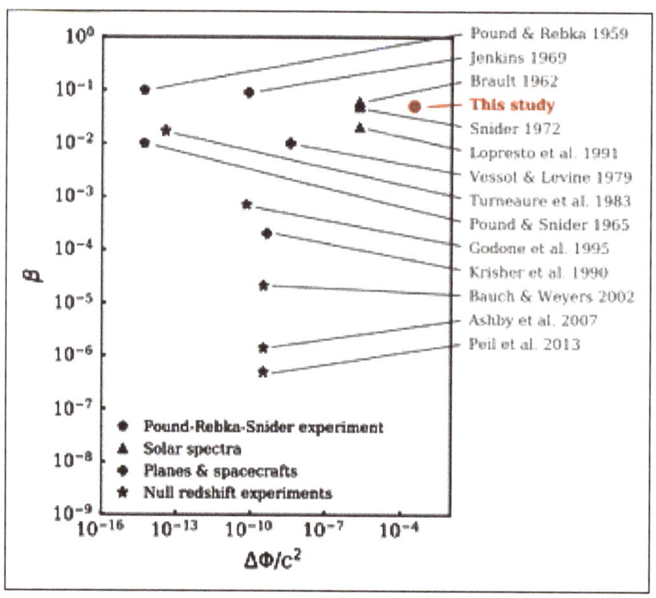

Comparison of different LPI studies with the S2 result shown in red.

What Stellar Orbit is Needed to Measure the Spin of Sgr A* from Astrometric Data?

The spin of a black hole causes the entire spacetime in its vicinity to be dragged along with the black hole rotation. This frame-dragging or Lens-Thirring effect is, however, a subtle quantity that affects the spacetime only in the immediate vicinity of the source of gravity. Therefore, in order to measure the spin of a black hole with stellar orbits, a star must have a trajectory that passes sufficiently close to the black hole. In the case of the Galactic centre massive black hole, Sgr A*, observed with GRAVITY/VLT, a star requires a certain combination of semi-major axis, aorb, and eccentricity, e, to enable a spin measurement based on a number of N observations within T years.

Spin measurement enabling stars are then expected on orbits with semi major axes of a several hundred Schwarzschild radii. Using the current distributions of eccentricities and semi major axes in the inner region of the Galactic centre, the estimated number of stars satisfying this conditions is about 0.12 for a 10y observing campaign. Including radial velocity measurements with precision 50 km/s helps as this provides stronger constraints on the parameters and increases the number of expected stars by a factor of 2.

Classical Tests

Perihelion Advance of Mercury

The first general relativistic effect, to be observed, was the advance of the perihelion in Mercury's orbit; it was viewed by Le Verrier in 1859. The advance of the perihelion sometimes referred to as apsidal precession or orbital precession, basically means the orbit is precessing. It would be extremely imprecise to measure the precession of one orbit, but measuring the accumulated perihelion advance over 100 years, makes the effect measurable. It is measured in arc seconds per century, and ascribed the following notation, $\Delta\phi$ century^{-1}.

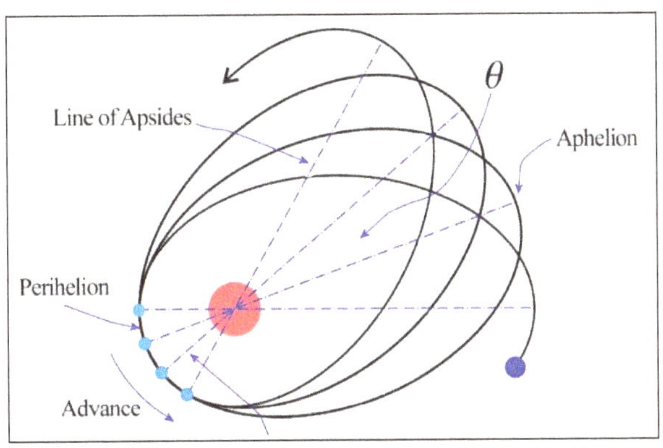

Precession of the perihelion, after one orbit makes the angle f. The effect is greatly exaggerated.

The Newtonian treatment for value for apsidal precession is concerned with forces, as is the law of universal gravitation. According to Newton's 'Principia,' Proposition 45 of Book I, the centripetal force could be represented as shown in Equation μr^{n-3}, for orbits almost circular; namely, the eccentricity e ≪ 1. The value of the constant m is often taken to be ~ 1. The index for the centripetal force n is a function of varying e. When there was no precession n = 1, then the value is quiescent, as the form resorts to the inverse square law. Hence, for apsidal precession $n \neq 1$. If e~ 0, then apsidal precession would not occur.

$$\mu r^{n-3}$$

The angle between lines of apsides is θ and is provided in equation; also note: $\theta \equiv \phi$.

$$\theta = \pi / \sqrt{n} = 180° / \sqrt{n}$$

In Newton's, 'System of the World,' Book 3, he states that precession can be accounted for by perturbation. The work does predict apsidal precession. However, following users of this method found that it failed to account fully for the phenomena.

Le Verrier used perturbation theory and planetary transit data, spanning 50 years. He calculated the interaction of Mercury with the other planets and found the residual of 39" century was not accounted for by Newtonian perturbations. In 1882, Newcomb determined the rate of precession to be 531" century^{-1}, leaving a residual of 43" century less than the observed rate, using similar methods. Theories emerged to resolve the residual, using Newtonian equations, but they were ad hoc and failed to predict much else. People speculated that the excess effect, or residual, was due to an unseen planet, they named Vulcan or some other collection of objects.

In 1915 Einstein derived the correct value of the precession of the perihelion of Mercury. The fact that other solar system planets have a significant gravitational effect on Mercury, albeit tiny, meant Mercury did not move in precisely a 1/r orbit, being the Newtonian potential. The GR equation of motion, or orbital equation, is provided in Equation below:

$$\frac{d^2u}{d\phi^2}+u=\frac{GM}{h^2}+\frac{3GM}{c^2}u^2$$

Using the notation of Hobson et al: the angular momentum, per unit mass, of the orbiting particle is $h,u \approx GM/h^2\{1+e\cos[\phi(1-\Delta u)]\}$, where Δu is the perturbation: $\Delta u = A \left[1+e^2\left(\frac{1}{2}-\frac{1}{6}\cos 2\phi\right)+e\phi\sin\phi\right]$ and $A=3(GM)^3/(c^2h^4)$, which is tiny. The residual $\Delta\phi$, can then be derived as shown in Equation $\Delta\phi=\frac{6\pi GM}{a(1-e^2)c^2}$, while the semi-major axis $a=5.8\times10^{10}\,\text{m}$, solar mass is $M_\odot=2\times10^{30}$ kg, and e = 0.2.

$$\Delta\phi=\frac{6\pi GM}{a(1-e^2)c^2}$$

Einstein's GR predicted value for $\Delta\phi$ was 43'' century^{-1}, while astronomers at that time determined $\Delta\phi$ to be, $45''\pm 5''$. Recently the observational value of $\Delta\phi$ has been constrained to 42."98 ± 04 century.

Deflection of Light

Einstein's work of 1907 also revealed that light was bent as a result of spacetime curvature. In the general relativistic case, this is only true of a stationary observer who sees the light pathway relative to a gravitating body. Einstein understood, using the EEP, that mass, or indeed energy from $E=mc^2$, would follow geodesic pathways in curved spacetime as depicted in Figure, relative to an observer at rest with the gravitating body.

This concept embodies the essence of EEP, showing that gravity and acceleration are indistinguishable from each other, within a small region. As the speed of light is constant, Einstein also showed that time dilation occurs in the reference frame where the light was produced. In part b), time has to run slower, or become dilated, to match the coordinate time of the reference frame in a). Meaning, light had to travel further without going greater than c.

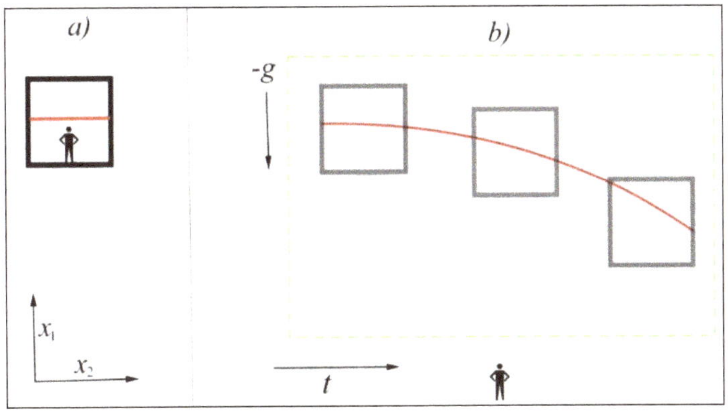

Part a) shows the reference frame that is at rest or in free fall, where light is being emitted. Part b) shows the same event, relative to an observer who sees the emitted light move as a result of gravity, in curved spacetime.

Assuming that light from a distant star is passing by the Sun or from any gravitating body, at its closest approach, Equation $\frac{d^2u}{d\delta^2} + u\frac{3GM}{R_0^2}\cos^2\delta$ represents the GR equation of motion of deflection.

$$\frac{d^2u}{d\delta^2} + u\frac{3GM}{R_0^2}\cos^2\delta$$

The radius of the Sun is, R_0, and $u \equiv 1/r$, being due solely to the Sun and is the Newtonian potential. The angle of deflection is d. Note the similarity with Equation $\frac{d^2u}{d\phi^2} + u = \frac{GM}{h^2} + \frac{3GM}{c^2}u^2$.

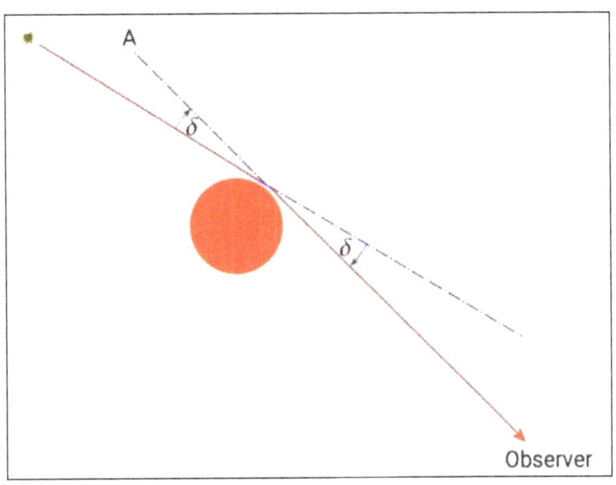

Figure: The bending of light by a massive object is, in this case, the Sun. Point A is the apparent location of the source as seen by the observer.

After some manipulation and a few assumptions, the net deflection of light according to GR δ_{GR} reduces to Equation $\delta_{GR} = \frac{4GM}{R_0c^2}$, as depicted in Figure.

$$\delta_{GR} = \frac{4GM}{R_0c^2}$$

In contrast, the Newtonian result for the deflection of light dNG is shown in Equation below. The results were such that, dNG resulted in $\sim 0''.87$ and dGR resulted in $\sim 1''.75$.

$$\delta_{\rightarrow} = \frac{2GM}{R_0 c^2} = \frac{1}{2}\delta_{GR}$$

Einstein suggested, in his 1911 paper, that this phenomenon may be testable during a solar eclipse. Then in 1917, Sir Frank Dyson the Astronomer Royal and Arthur Eddington planned to photograph a total solar eclipse on May 29, 1919, as the sun was crossing the Hyades cluster. Eddington measured the true positions of Hyades between January and February of 1919. Then in May 1919, he travelled to the island of Príncipe, off the west coast of Africa, situated in the Gulf of Guinea. There, during that total solar eclipse, he measured the apparent change in position, or displacement, of the stars in the Hyades cluster as a result of the Sun's gravitational lens. The stars nearest the Sun's rim appeared to move outwards radially from the Sun as depicted in Figure.

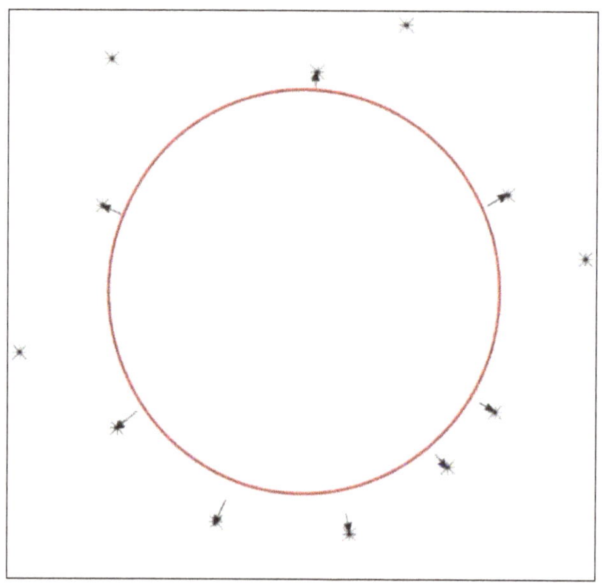

When photons from stars are passing close by the Sun, as in a solar eclipse, their apparent position seems to move radially outward.

On November 6th, 1919 Eddington presented his findings. The results of two different photographic plates of the displacement were stated: 1".98 ± 0".12 and 1".61 ± 0".30. Eddington said that the predictions of GR were compatible with the results of their observed deflection, whereas the Newtonian was not. The theoretical prediction of GR was 1".75. However, there are other factors to consider.

In succeeding years, more measurements were made, but there was no significant improvement until the advent of very-long-baseline interferometry [VLBI] in 1972. VLBI is the simultaneous use of various radio telescopes that are situated at distant locations from one another which act as an interferometer. This technique creates a virtual radio telescope with a far larger baseline. VLBI uses a geometric method to measure the arrival time of radio waves from an astronomical source between different telescopes τ, and their angular separation β and a geometric measure of the projected baseline U. The VLBI principle is shown in Figure with two radio telescopes. The signal delay L is a trigonometric relationship between those variables and the baseline B.

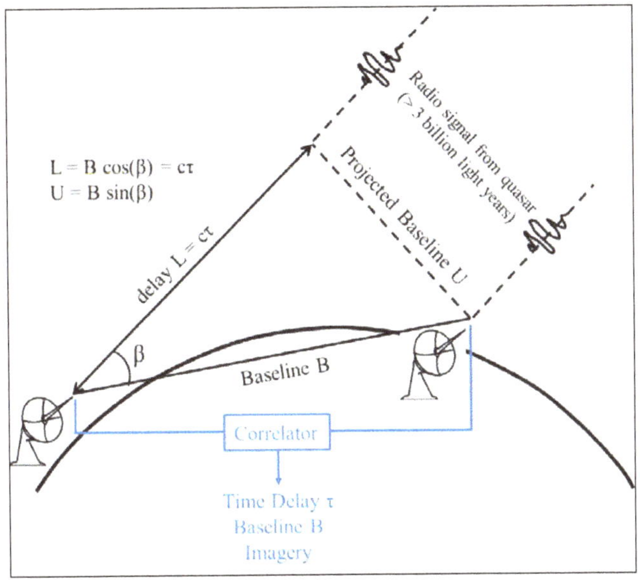

VLBI detecting radio signals from a quasar.

Shapiro et al., reports on deflection by the Sun of radio waves emitted from extragalactic radio sources, between 1979 and 1999. Using VLBI for radio band detection provides a greater level of accuracy over the visible band, for the visible band is degraded by atmospheric absorption and scintillation. The PPN [defined p. 8] framework results were, $\gamma = 0.9998 \pm 0.0004$, compared to the GR value of, $\gamma = 1$.

These tests of the bending of light, or gravitational lensing, are observational evidence that photons follow geodesics in spacetime, being a fundamental precept in GR. This was an early monumental verification for GR.

Gravitational Redshift

In 1925 Adams measured the gravitational redshift zg, of Sirius B, which was non-Doppler and measured to be $zg = +23\,\text{km}^{s\text{-}1}$, were not stated. This was the first observational verification of gravitational redshift [GvR], yet Adams's measurement was only one quarter the actual value. More recently, the Hubble Space Telescope [HST], measured the value of GvR of Sirius B to be, $zg = +\,80.42 \pm 4.83\,\text{kms}^{-1}$.

In 1959, Pound and Rebka, devised a precision method to experimentally measure the GvR, using the Mössbauer effect. When a photon is emitted or absorbed, an exchange of momentum with the nuclei occurs, recoiling the nucleus which reduces the emitted photon's net energy, hence its frequency. This is a result of the conservation of momentum. Rudolf Mössbauer in 1958, determined that an almost recoil-free emission and absorption of gamma radiation can occur by embedding atomic nuclei in a solid lattice crystal so that it was resistant to recoil. This is the Mössbauer effect and would permit researchers to detect the GvR effect without contamination by the recoil of the nucleus during absorption or emission, and reduced energy of emitted photons.

In the following year, 1960, Pound and Rebka arranged their experiment in a high tower at Jefferson laboratory, on the Harvard campus. The physical setup is depicted in Figure. The height of the experiment is h, being 22.6 m. The fractional width and intensity of Fe spectral lines seemed sufficient to measure the GvR effect in the laboratory and was therefore selected. The source excited Fe ion which released gamma photons was embedded in a lattice.

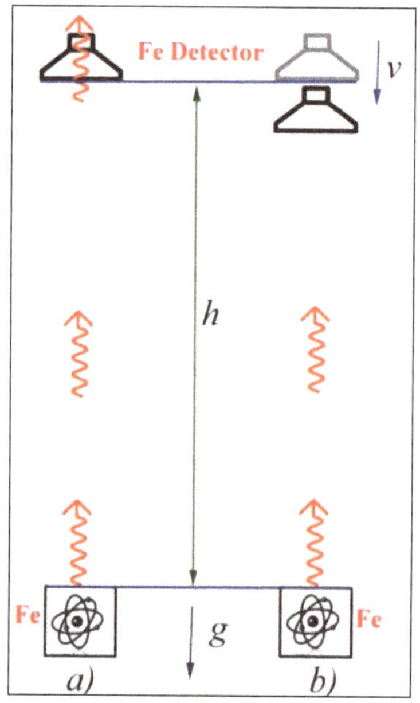

This is an illustration of the Pound and Rebka experiment when in actual fact the experiment was also conducted in an inverted configuration. In part a) the detector is stationary and the photons pass right through. In part b) the detector is moving with velocity v and the photons are absorbed.

As the photon climbed upward against the gravitational field of the Earth g, the photon became gravitationally redshifted. When the photon arrives at the Fe ion detector, it has a lower frequency as seen in part a), and was not absorbed. In part b), the Fe ion detector is moved towards the emitter; when the Doppler blueshift of the moving emitter is the same in magnitude, but opposite in sign, as the GvR effect; the photon has the right energy and is absorbed.

To an observer in b) at rest, the falling detector has increased its speed, as in Equation $g\Delta t = gh/c$.

$$g\Delta t = gh/c$$

Hence, the velocity of the detector is, $v = g\Delta t$. when travelling towards the Earth the effect is a blueshifted Doppler shift which is depicted by Equation $\Delta\lambda = \lambda(v/c) = \lambda(gh/c^2)$.

$$\Delta\lambda = \lambda(v/c) = \lambda(gh/c^2)$$

Therefore, the gravitational redshift z_g, can then be represented by Equation 18, where the gravitational potential of the Earth is FE. Φ_E.

$$z_g = \frac{gh}{c^2} = \frac{\Delta\Phi_E}{c^2}$$

This experiment was repeated many times and achieved the first physical verification of GvR, measuring a net fractional shift of ~ – (5.13 ± 0.51) × 10-15. In contrast to the GR estimated value of, ~ – (4.92 ± 0.51) × 10^{-15}. The precision of this measurement was independent of height, while later versions of the experiment constrained the result to within ~ 1% of the GR prediction.

This result was a huge scientific achievement for experimental testing of GR, not only because it used the Mössbauer effect and produced a high-resolution result, but also because of the unique and ingenious experimental arrangement. The accuracy of these tests is comparable to the precision of modern ones.

Many experimental tests of GvR followed, while some used hydrogen masers in space. A Scout rocket was launched in 1976, which carried the Vessot–Levine Experiment as part of Gravity Probe A [GP-A]. GP-A was a NASA a Space-Based geodetic satellite to test GR. This experiment constrained the relative discrepancy with GR to ~70 × 10^{-4} level. The Russian satellite RadioAstron reduced the relative discrepancy to ~10^{-6}. These latter results have constrained GvR to high levels of precision. Lastly, the tests of GvR are classified as the third classical test of GR.

Shapiro Delay

The light travel time delay, or Shapiro delay, is sometimes called the fourth classical test of GR and was first introduced by Shapiro in 1964. This delay in the arrival time of light passing nearby a massive object is the casual result of general relativistic time dilation, due to a significant gravitational potential. Shapiro realised, the speed of light is dependent on the gravitational potential along its travel path, according to GR, and could be tested with a close conjunction of the Sun. These tests were measured in the weak-field regime. The physical definition of r_p, r_e and r_0, during a test scenario are represented by figure below.

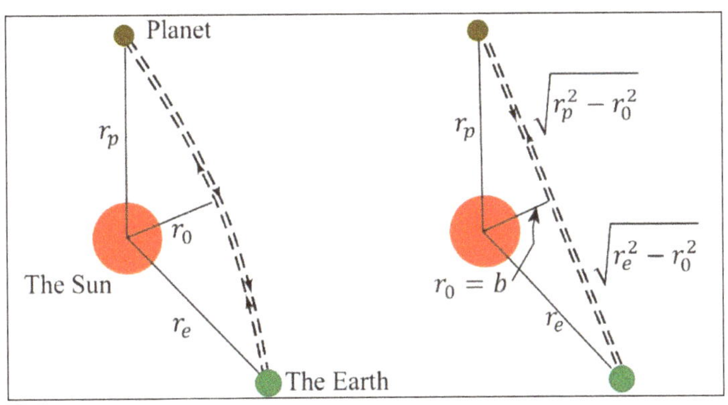

The radar reflection of photons from the Earth to a planet and back. The left image is the actual path, exaggerated. The right image is the Euclidean form.

To define Shapiro delay, assume the Earth and the planet are stationary, while the total time for the round trip of the radar signal is Δt, in coordinate time. The value of t must be represented in terms of r along the entire pathway, while r_0 is the closest approach to the Sun or its radius. The travel time from Earth to the reflecting planet and back to Earth is given in Equation $\Delta t \approx \frac{2}{c}\left[\left(r_e^2 + r_0^2\right)^{1/2} + \left(r_p^2 + r_0^2\right)^{1/2}\right] + \frac{4k}{c}\left\{\text{In}\left(4\frac{r_e r_p}{r_0^2}\right) + 1\right\}$, where k is the Schwarzschild metric parameter: $k = GM_{\odot} / c^2$. Also r_e s the orbital radius of the Earth and r_p is the orbital radius of the reflecting planet.

$$\Delta t \approx \frac{2}{c}\left[\left(r_e^2 + r_0^2\right)^{1/2} + \left(r_p^2 + r_0^2\right)^{1/2}\right] + \frac{4k}{c}\left\{\text{In}\left(4\frac{r_e r_p}{r_0^2}\right) + 1\right\}$$

This leads to a viable solution for the excess time delay, or Shapiro delay Δtexcess, provided in Equation $\Delta t_{excess} \approx \frac{4GM}{c^3}\left[\ln\left(\frac{r_p r_e}{r_0}\right)+1\right]$.

$$\Delta t_{excess} \approx \frac{4GM}{c^3}\left[\ln\left(\frac{r_p r_e}{r_0}\right)+1\right]$$

In 1965, Shapiro et al., using the Lincoln Laboratory Haystack radar system, began a program to measure the delay in reflected radar signals off solar system planets. The radar system had ~ 200 msec accuracy. In 1966, they measured the Shapiro delay of signals bounced off Venus, and also from Mercury in 1967, both during a super-conjunction when the Shapiro delay signal is the strongest as the gravitational potential is highest. The time delay corresponded with GR predicted value to within, ~ 20%.

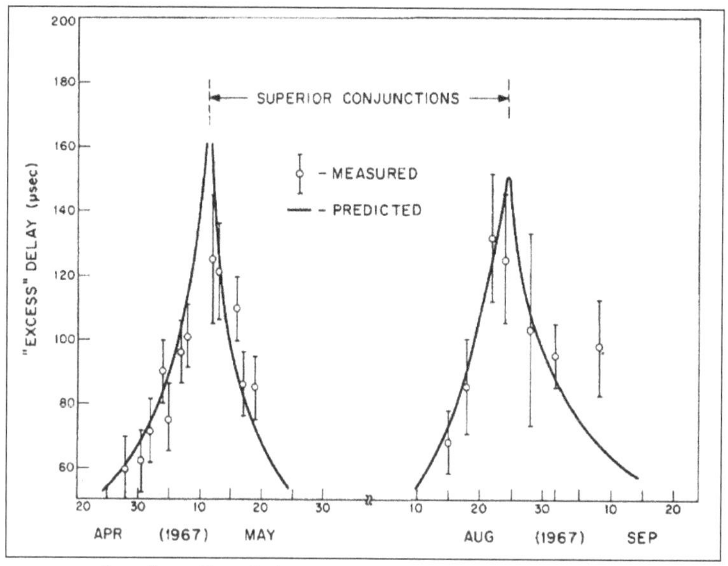

Figure shows measured and predicted Shapiro delays, in msec, based on GR for Earth-Mercury.

Subsequent measurements in 1971 further reduced the deviation from GR to ~ 5%. Then in 1979, transponders were placed on the Viking spacecraft, whereby radar signals were bounced off Mars from the Viking lander. This enabled Shapiro and Reasenberg to improve the accuracy of the Shapiro delay results, to better than ~ 0.1% with GR.

In 2003, the Cassini spacecraft flew by Saturn and measured the Shapiro time delay of bounced signals. The deviation of the PPN parameter γ, being the amount of curvature produced per unit mass, was 1.000021 ± 0.000023, against the GR value of $\gamma = 1$ (Ni, 2005). This result confirmed the GR effect of Shapiro delay to approximately 20 parts per million. Fundamentally, Shapiro time delay was not originally thought of by Einstein and his contemporaries, and so its later introduction and experimental verification elucidated GR.

Gravitational Time Dilation

Gravitational time dilation [GvT] is distinctly different from time dilation in special relativity [SR] and also cosmological time dilation. Equation only provides the first order of GvT, whereas the exact form is shown in Equation $d\tau = \left(1 + \frac{2\Phi}{c^2}\right)^{1/2} dt$. This provides the small element of proper time dt, being dilated from the coordinate time $d\tau$, near a massive object.

$$d\tau = \left(1 + \frac{2\Phi}{c^2}\right)^{1/2} dt$$

Expanding this relation, with the potential Φ depicted in Equation 1; namely $\Phi = -GM/r$ provides Equation $d\tau = \left(1 - \frac{2GM}{rc^2}\right)^{1/2} dt$. This asserts that when $\Phi \to 0$, then $d\tau \to dt$. hence, when there is no gravitational potential, there is no time dilation.

$$d\tau = \left(1 - \frac{2GM}{rc^2}\right)^{1/2} dt$$

The first direct experimental confirmation of GvT was performed in 1971 by Hafele and Keating, who flew four Cs-clocks around the world, both eastward and westward. Clocks in the plane lost −59 ± 10 ns travelling eastward, and gained + 273 ± 7 ns travelling westward. The respective GR predictions were −40 ± 23 ns; + 275 ± 21 ns. This too, being in Earth's orbit was measured in the weak-field regime. The resulting total time dilation was in agreement with SR and GR, to within $\sim 10\%$. In the Hafele and Keating experiment, the total time dilation t, for one trip, is represented by equation, where to is the initial time, and the altitude is h.

$$\tau = \tau_0\left[gh/c^2 - \left(2R\Omega\mathrm{v} + \mathrm{v}^2\right)/2c^2\right]\tau_0$$

A stationary equatorial clock on Earth, relative to non-rotating space, has a 'clock rate' of $R\Omega$, while v is the velocity, gh is the gravitational potential, and gh/c^2 is the GvR term. The results are shown in Figure. Many subsequent tests of this kind, have been carried out, further reducing the uncertainty to $\sim 1.6\%$.

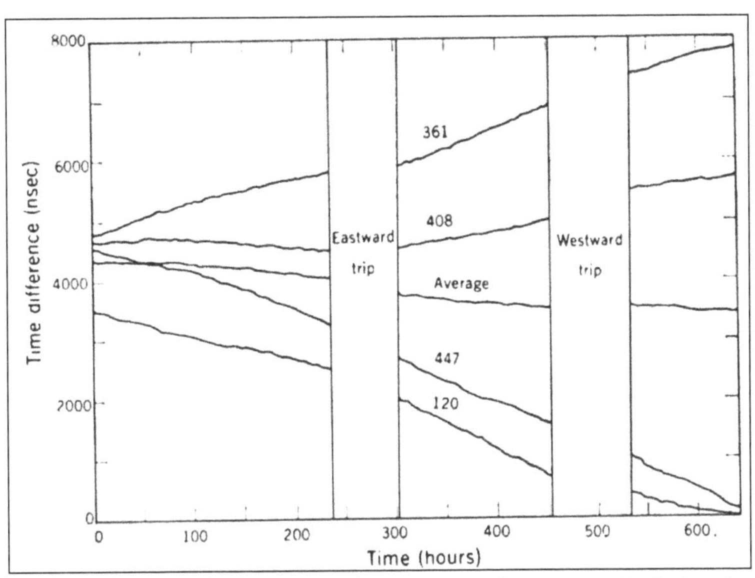

Hafele and Keating experiment, showing the time differences between each clock at hourly intervals.

A similar test was performed, with the first FRG Spacelab Mission Dl, on the Space Shuttle Challenger in 1985. The on-board NAVEX Cs-clock, against the ground reference clock, determined GvT to $\sim 5\ \times 10^{-4}$, with an 1σ uncertainty.

The Global Positioning System [GPS] basically consists of 24 satellites with on-board atomic clocks, is a daily validation of SR and GR. GPS calculations utilize the Kerr metric, The GPS satellite system runs on Universal Coordinated Time [UTC], and is managed by the U.S. Naval Observatory [USNO]. The determination of transfer time and position is represented by Equation $c^2\left(t-t_j\right)^2 = \left|r-r_j\right|^2$. For example, using n synchronized atomic clocks, transmitting pulses from their positions at distances r_j, and at times t_j, where j = 1, 2,..n. These signals are received, at a given satellite, at a distance from the Earth by the position vector r, at time t, due to the constancy of the speed of light, the following relation holds.

$$c^2\left(t-t_j\right)^2 = \left|r-r_j\right|^2$$

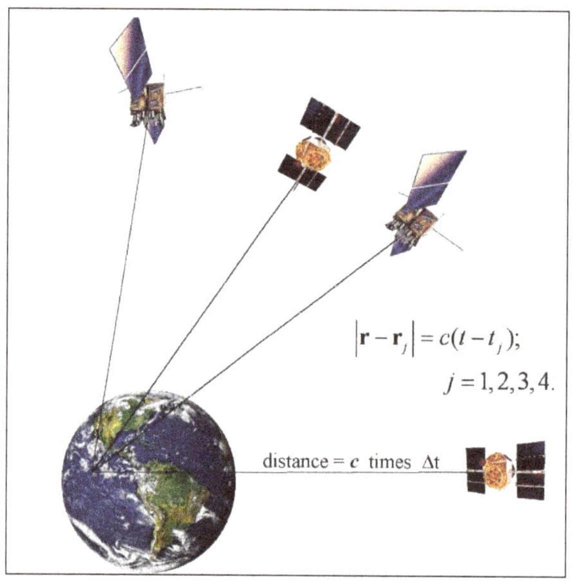

The positions of receivers are shown. Transmitted at t_j, and at positions r_j.
The values of GvT are found by solving four simultaneous equations.

The GPS has various relativistic effects; namely, time dilation due to SR, GvT, and GvR, as well as frame-dragging, and the Sagnac effect. However, our interest here is purely the GvT value due to the GR, according to Zhao et al., this value is, $\sim 4.5 \times 10^{-10}$, or 4.5 in 10^{10} parts, being in the weak-field regime.

Frame-Dragging and Geodetic Effect

Other phenomena that are predicted by GR are the 'frame–dragging' of spacetime or Lense– Thirring precession, and the 'geodetic effect.' These are discussed respectively. These tests utilize the Kerr metric, as does the GPS.

Lense–Thirring Precession

Lense–Thirring precession [LTP] is distinctly different from de Sitter precession; having been tested successfully marks an important step in GR's confirmation. The dragging of inertial frames by a massive body is a direct consequence of GR, although, it was Lense and Thirring in 1918, who derived the equations of frame–dragging, using the Kerr metric.

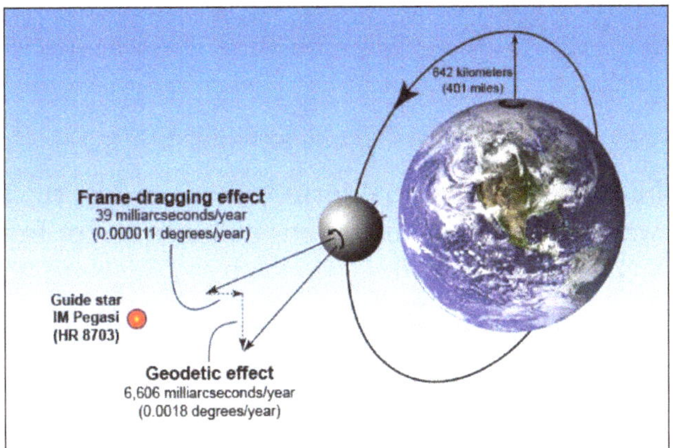

	Measured	Predicted
Geodetic precession (mas)	6602 ± 18	6606
Frame-dragging (mas)	37.2 ± 7.2	39.2

1) Frame-dragging, and 2) the Geodetic effect.

LTP can be seen in a satellite in a polar orbit around the Earth, as the Earth undergoes precession about its axis, through the poles. LTP predicts that a satellite's plane of orbit will precess slowly in the direction of the Earth's rotation. The effect is depicted in Figure above.

LTP is represented by Equation $\Omega = \frac{2GI\omega}{c^2a^3\left(1-e^2\right)^{3/2}}n^2$. The value I is the moment of inertia and ω is the angular velocity, both of the central mass: the Earth. Also, a and e are the semi-major axis and eccentricity of the orbiting body, respectively. The unit vector n is parallel to the central body's spin axis. Hence, the resultant is the advance rate of precession W.

$$\Omega = \frac{2GI\omega}{c^2a^3\left(1-e^2\right)^{3/2}}n^2$$

Ciufolini and Pavlis in 2004, reported measuring LTP on the LAGEOS30, LAGEOS2, and GRACE to 99 ± 5% of that predicted by GR, with a 10% overall uncertainty. This uncertainty was later reduced to 0.994 × 0.002 to the value of GR.

Geodetic Precession

The next phenomenon is the geodetic effect, also known as geodetic precession [GP], which is the product of the GR's spacetime curvature, and was measured close to the Earth as a result of the Earth's mass. Figure, shows the GP in relation to the LTP.

In 1960, Schiff demonstrated that an ideal gyroscope in Earth's orbit, being in the weak-field regime, would experience two relativistic effects: GP and LTP. Schiff's formula is provided in Equation $\Omega = \frac{3GM}{2c^2R^3}(R \times v) + \frac{GI}{c^2R^3}\left[\frac{3R}{R^2}(\omega.R) - \omega\right]$. Moreover, the GP moves in the direction of the orbit, being orthogonal to LTP.

$$\Omega = \frac{3GM}{c^2R^3}(R \times v) + \frac{GI}{c^2R^3}\left[\frac{3R}{R^2}(\omega R) - \omega\right]$$

On the right-hand side of Equation 26, the first term is the GP while the second term is a reworking of Equation 25: LTP, where R is the instantaneous distance vector from the Earth, and v is the velocity of the gyroscope. GP-B measured the GP to $\sim 0.5\%$ against GR. Over longer periods of operation, this result was improved to $\sim 0.28\%$.

Utilizing Lunar Laser Ranging [LLR] as a means of testing GR provided the value of GP expressed as a deviation to -0.0019 ± 0.0064, with GR being 0.

Equivalence Principle Tests

The weak equivalence principle [WEP] is the equality of gravitational and inertial mass, so that $M_G \equiv M_I$, while Einstein's equivalence principle [EEP] is the equivalence of homogeneous gravity and uniform acceleration in small regions, depicted in Figure. However, EEP is dependent on the WEP also being valid. Tests of the WEP are generally concerned with testing the universality of free fall [UFF]; namely, that all masses in a uniform gravitational field, experience the same acceleration.

Some authors dichotomize EEP into two classes, first is EEP and the second is the strong equivalence principle [SEP]. These imply two significant results for locally non-gravitational reference frames:

- Any experiment is independent of the velocity of the UFF reference frame, referred to as local Lorentz invariance [LLI].
- Any experiment is independent of position and time in the universe; this is local position invariance [LPI].

If these two statements are valid, then SEP and EEP are correct. Moreover, SEP contains both EEP and the WEP. However, SEP includes the gravitational self-energy of bodies, whereas EEP does not. Hence, SEP requires the involvement of astrophysical objects. Experiments of EEP have a negligible gravitational self-energy. There have been many tests of the WEP, but this report is only concerned with tests of EEP and the SEP.

Tests of EEP and SEP usually employ the PPN framework and also the Equivalence Principle parameter η, which is a function of the PPN parameters β and γ, which are often called the Eddington–Robertson–Schiff parameters. The simplest relationship, using only γ and β, is in Equation $\eta = 4\beta - \gamma - 3$, often called the Nordtvedt equation; in the weak-field regime.

$$\eta = 4\beta - \gamma - 3$$

If EEP and SEP is 100% correct, then $\eta = 0$. Generally, SEP or EEP can be validated by the presence of GvT, or Shapiro time delay, or GvR. Early suggestions of EEP tests were reported in the 1960s.

In 1979, a hydrogen maser in a space probe at an altitude of 10,000 m was used to measure the GvT, confirmed EEP to $\eta = 2 \times 10^{-4}$.

The supernova SN1987A provided a stringent test of EEP on intergalactic scales, by the almost simultaneous arrival of the photons and neutrinos. This result was in agreement with the GR prediction to ~ 0.2%. In 2003, data from the Cassini spacecraft, using Shapiro time delay, measured $\gamma = 1.000021 \pm 0.000023$,. Spectral distortions of the Cosmic Microwave Background radiation [CMB], using COBE/FIRAS, indicate a deviation from EEP no greater than, < 10^{-5} within CMB energy scales. EEP has been thoroughly tested.

Suggestions of various methods to verify the SEP also started in the early 1960s. LLR has been crucial for testing GR. These tests have provided stringent tests of the SEP and the time variation of G.

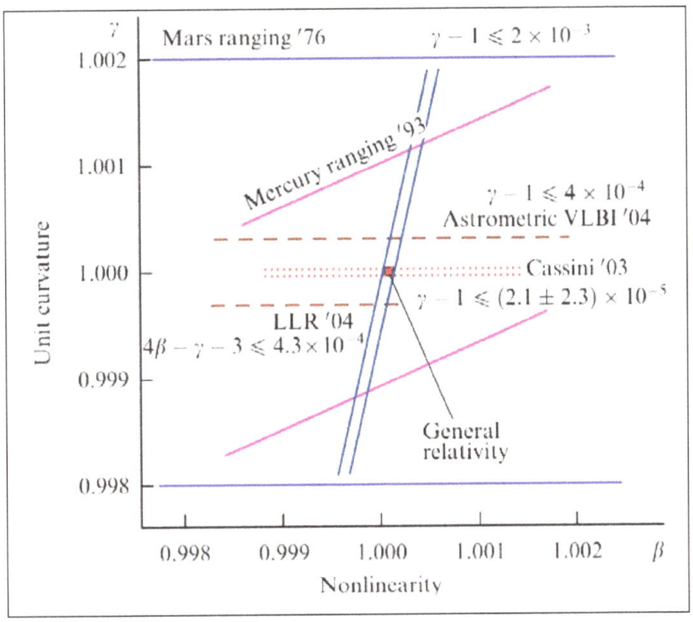

Values of γ plotted against b from 1969 to 2009, yielding the best values of $\gamma - 1 = (2.1 \pm 2.3) \times 10^{-5}$ and $\beta - 1 = (1.2 \pm 1.1) \times 10^{-4}$.

LLR researchers were able to test the SEP, using the prediction that gravitational binding energy falls like other forms of mass and energy relations. Over 35 years of LLR tests resulted in $\eta = (2.3 \pm 3.2) \times 10^{-4}$, against the GR prediction of $\eta = 0$. Recent ground and LLR experiments have yielded a strict upper limit on deviations from the SEP, where $\eta = 4.4 \times 10^{-4}$. Figure, depicts the improvements in determining the Eddington parameters γ and β, within the weak-field regime.

Testing the variation in time of fundamental constants are also methods for determining if the SEP is violated. Generally, constraining the value of η for EEP is a good test for the weak-field regime within the solar system but it can also be employed for the SEP on cosmological scales.

Solar-System Tests

Many of the most important tests of GR have been performed in the solar system, and some have been covered in previous sections of this report. Measurements within the solar system are concerned with the weak-field regime and are considered asymptotically flat in GR.

Moreover, solar system tests have confirmed GR to a high level of accuracy (Ni, 2016). Whilst there have been many solar system studies, and many planned for the future, only a couple notable ones are briefly mentioned.

Nordtvedt, using solar system observations, stated that there was less than 1 part in 10^9 of a departure from GR. Anderson et al., using the Jet Propulsion Laboratory [JPL] ephemerides data of planetary positions, together with contemporary data analysis methods and PPN parameters, obtained the following. Planetary positions were mapped at different times, it was found that, $(\beta-1)=-0.0010\pm0.0012$, and $(\gamma-1)=(0-0.0015)\pm0.0021$, whereby the GR predictions are, $(\beta-1)=(\gamma-1)=0$. EEP and the SEP haven been extensively tested within the weak-field regime.

References

- Theory-special-relativity-36273: space.com, Retrieved 10, April 2020
- The-lorentz-transformation, universityphysicsv3openstax: opentextbc.ca, Retrieved 29, March 2020
- What-general-relativity, content: plus.maths.org, Retrieved 11, July 2020
- Einstein-field-equation, physics: byjus.com, Retrieved 19, August 2020
- Principle-of-covariance-77: einsteinrelativelyeasy.com, Retrieved 14, June 2020
- Tests-of-General-Relativity-7260308: mpe.mpg.de, Retrieved 23, February 2020

4

A Comprehensive Study of Physical Cosmology

Physical cosmology is a branch of cosmology concerned with dynamics of the universe as well as the origin, structure, evolution and ultimate fate of the universe. This chapter discusses physical cosmology, Friedmann-Lemaitre-Robertson-Walker Metric, big-bang nucleosynthesis and absolute time and space. All the diverse principles of physical cosmology have been carefully analyzed in this chapter.

Physical Cosmology

Physical cosmology is the study of the largest-scale structures and dynamics of the Universe and is concerned with fundamental questions about its origin, structure, evolution, and ultimate fate. Cosmology as a science originated with the Copernican principle, which implies that celestial bodies obey identical physical laws to those on Earth, and Newtonian mechanics, which first allowed us to understand those physical laws. Physical cosmology, as it is now understood, began with the development in 1915 of Albert Einstein's general theory of relativity, followed by major observational discoveries in the 1920s: first, Edwin Hubble discovered that the universe contains a huge number of external galaxies beyond our own Milky Way; then, work by Vesto Slipher and others showed that the universe is expanding. These advances made it possible to speculate about the origin of the universe, and allowed the establishment of the Big Bang Theory, by Georges Lemaitre, as the leading cosmological model. A few researchers still advocate a handful of alternative cosmologies; however, most cosmologists agree that the Big Bang theory explains the observations better.

Dramatic advances in observational cosmology since the 1990s, including the cosmic microwave background, distant supernovae and galaxy redshift surveys, have led to the development of a standard model of cosmology. This model requires the universe to contain large amounts of dark matter and dark energy whose nature is currently not well understood, but the model gives detailed predictions that are in excellent agreement with many diverse observations.

Cosmology draws heavily on the work of many disparate areas of research in theoretical and applied physics. Areas relevant to cosmology include particle physics experiments and theory, theoretical and observational astrophysics, general relativity, quantum mechanics, and plasma physics.

Subject History

Modern cosmology developed along tandem tracks of theory and observation. In 1916, Albert Einstein published his theory of general relativity, which provided a unified description of gravity as a geometric property of space and time. At the time, Einstein believed in a static universe, but found that his original formulation of the theory did not permit it. This is because masses distributed throughout the universe gravitationally attract, and move toward each other over time. However, he realized that his equations permitted the introduction of a constant term which could counteract the attractive force of gravity on the cosmic scale. Einstein published his first paper on relativistic cosmology in 1917, in which he added this *cosmological constant* to his field equations in order to force them to model a static universe. However, this so-called Einstein model is unstable to small perturbations—it will eventually start to expand or contract. The Einstein model describes a static universe; space is finite and unbounded (analogous to the surface of a sphere, which has a finite area but no edges). It was later realized that Einstein's model was just one of a larger set of possibilities, all of which were consistent with general relativity and the cosmological principle. The cosmological solutions of general relativity were found by Alexander Friedmann in the early 1920s. His equations describe the Friedmann–Lemaître–Robertson–Walker universe, which may expand or contract, and whose geometry may be open, flat, or closed.

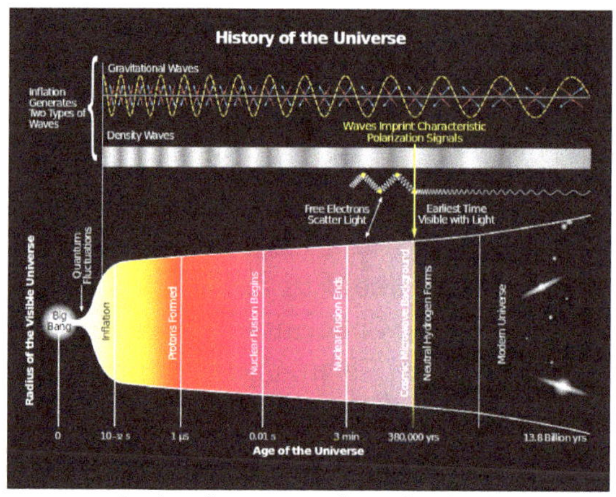

History of the Universe – gravitational waves are hypothesized to arise from cosmic inflation, a faster-than-light expansion just after the Big Bang.

In the 1910s, Vesto Slipher (and later Carl Wilhelm Wirtz) interpreted the red shift of spiral nebulae as a Doppler shift that indicated they were receding from Earth. However, it is difficult to determine the distance to astronomical objects. One way is to compare the physical size of an object to its angular size, but a physical size must be assumed to do this. Another method is to measure the brightness of an object and assume an intrinsic luminosity, from which the distance may be determined using the inverse square law. Due to the difficulty of using these methods, they did not realize that the nebulae were actually galaxies outside our own Milky Way, nor did they speculate about the cosmological implications. In 1927, the Belgian Roman Catholic priest Georges Lemaître independently derived the Friedmann–Lemaître–Robertson–Walker equations and proposed, on the basis of the recession of spiral nebulae, that the universe began with the "explosion" of a "primeval atom"—which was later called the Big Bang. In 1929, Edwin Hubble provided an observational basis for Lemaître's theory. Hubble showed that the spiral nebulae were galaxies by determining their distances using measurements of the brightness of Cepheid variable stars. He discovered a relationship between the redshift of a galaxy and its distance. He interpreted this as evidence that the galaxies are receding from Earth in every direction at speeds proportional to their distance. This fact is now known as Hubble's law, though the numerical factor Hubble found relating recessional velocity and distance was off by a factor of ten, due to not knowing about the types of Cepheid variables.

Given the cosmological principle, Hubble's law suggested that the universe was expanding. Two primary explanations were proposed for the expansion. One was Lemaître's Big Bang theory, advocated and developed by George Gamow. The other explanation was Fred Hoyle's steady state model in which new matter is created as the galaxies move away from each other. In this model, the universe is roughly the same at any point in time.

For a number of years, support for these theories was evenly divided. However, the observational evidence began to support the idea that the universe evolved from a hot dense state. The discovery of the cosmic microwave background in 1965 lent strong support to the Big Bang model, and since the precise measurements of the cosmic microwave background by the Cosmic Background Explorer in the early 1990s, few cosmologists have seriously proposed other theories of the origin and evolution of the cosmos. One consequence of this is that in standard general relativity, the universe began with a singularity, as demonstrated by Roger Penrose and Stephen Hawking in the 1960s.

An alternative view to extend the Big Bang model, suggesting the universe had no beginning or singularity and the age of the universe is infinite, has been presented.

Energy of the Cosmos

Light chemical elements, primarily hydrogen and helium, were created in the Big Bang process. The small atomic nuclei combined into larger atomic nuclei to form heavier elements such as iron and nickel, which are more stable. This caused a *later energy release*. Such reactions of nuclear particles inside stars continue to contribute to *sudden energy releases*, such as in nova stars. Gravitational collapse of matter into black holes is also thought to power the most energetic processes, generally seen at the centers of galaxies.

Cosmologists cannot explain all cosmic phenomena exactly, such as those related to the accelerating expansion of the universe, using conventional forms of energy. Instead, cosmologists propose a new form of energy called dark energy that permeates all space. One hypothesis is that dark energy is the energy of virtual particles, which are believed to exist in a vacuum due to the uncertainty principle.

There is no clear way to define the total energy in the universe using the most widely accepted theory of gravity, general relativity. Therefore, it remains controversial whether the total energy is conserved in an expanding universe. For instance, each photon that travels through intergalactic space loses energy due to the redshift effect. This energy is not obviously transferred to any other system, so seems to be permanently lost. On the other hand, some cosmologists insist that energy is conserved in some sense; this follows the law of conservation of energy.

Thermodynamics of the universe is a field of study that explores which form of energy dominates the cosmos – relativistic particles which are referred to as radiation, or non-relativistic particles referred to as matter. Relativistic particles are particles whose rest mass is zero or negligible compared to their kinetic energy, and so move at the speed of light or very close to it; non-relativistic particles have much higher rest mass than their energy and so move much slower than the speed of light.

As the universe expands, both matter and radiation in it become diluted. However, the energy densities of radiation and matter dilute at different rates. As a particular volume expands, mass energy density is changed only by the increase in volume, but the energy density of radiation is changed both by the increase in volume and by the increase in the wavelength of the photons that make it up. Thus the energy of radiation becomes a smaller part of the universe's total energy than that of matter as it expands. The very early universe is said to have been 'radiation dominated' and radiation controlled the deceleration of expansion. Later, as the average energy per photon becomes roughly 10 eV and lower, matter dictates the rate of deceleration and the universe is said to be 'matter dominated'. The intermediate case is not treated well analytically. As the expansion of the universe continues, matter dilutes even further and the cosmological constant becomes dominant, leading to an acceleration in the universe's expansion.

History of the Universe

The history of the universe is a central issue in cosmology. The history of the universe is divided into different periods called epochs, according to the dominant forces and processes in each period. The standard cosmological model is known as the Lambda-CDM model.

Equations of Motion

The equations of motion governing the universe as a whole are derived from general relativity with a small, positive cosmological constant. The solution is an expanding universe; due to this expansion, the radiation and matter in the universe cool down and become diluted. At first, the expansion is slowed down by gravitation attracting the radiation and matter in the universe. However, as these become diluted, the cosmological constant becomes more dominant and the expansion of the universe starts to accelerate rather than decelerate. In our universe this happened billions of years ago.

Particle Physics in Cosmology

Particle physics is important to the behavior of the early universe, because the early universe was so hot that the average energy density was very high. Because of this, scattering processes and decay of unstable particles are important in cosmology.

As a rule of thumb, a scattering or a decay process is cosmologically important in a certain cosmological epoch if the time scale describing that process is smaller than, or comparable to, the time scale of the expansion of the universe. The time scale that describes the expansion of the universe is $1/H$ with H being the Hubble constant, which itself actually varies with time. The expansion timescale $1/H$ is roughly equal to the age of the universe at that time.

Timeline of the Big Bang

Observations suggest that the universe began around 13.8 billion years ago. Since then, the evolution of the universe has passed through three phases. The very early universe, which is still poorly understood, was the split second in which the universe was so hot that particles had energies higher than those currently accessible in particle accelerators on Earth. Therefore, while the basic features of this epoch have been worked out in the Big Bang theory, the details are largely based on educated guesses. Following this, in the early universe, the evolution of the universe proceeded according to known high energy physics. This is when the first protons, electrons and neutrons formed, then nuclei and finally atoms. With the formation of neutral hydrogen, the cosmic microwave background was emitted. Finally, the epoch of structure formation began, when matter started to aggregate into the first stars and quasars, and ultimately galaxies, clusters of galaxies and superclusters formed. The future of the universe is not yet firmly known, but according to the ΛCDM model it will continue expanding forever.

Areas of Study

Below, some of the most active areas of inquiry in cosmology are described, in roughly chronological order. This does not include all of the Big Bang cosmology, which is presented in *Timeline of the Big Bang*.

Very Early Universe

The early, hot universe appears to be well explained by the Big Bang from roughly 10^{-33} seconds onwards, but there are several problems. One is that there is no compelling reason, using current particle physics, for the universe to be flat, homogeneous, and isotropic. Moreover, grand unified theories of particle physics suggest that there should be magnetic monopoles in the universe, which have not been found. These problems are resolved by a brief period of cosmic inflation, which drives the universe to flatness, smooths out anisotropies and inhomogeneities to the observed level, and exponentially dilutes the monopoles. The physical model behind cosmic inflation is extremely simple, but it has not yet been confirmed by particle physics, and there are difficult problems reconciling inflation and quantum field theory. Some cosmologists think that string theory and brane cosmology will provide an alternative to inflation.

Another major problem in cosmology is what caused the universe to contain far more matter than antimatter. Cosmologists can observationally deduce that the universe is not split into regions of matter and antimatter. If it were, there would be X-rays and gamma rays produced as a result of annihilation, but this is not observed. Therefore, some process in the early universe must have created a small excess of matter over antimatter, and this (currently not understood) process is called *baryogenesis*. Three required conditions for baryogenesis were derived by Andrei Sakharov in 1967, and requires a violation of the particle physics symmetry, called CP-symmetry, between matter and antimatter. However, particle accelerators measure too small a violation of CP-symmetry to account for the baryon asymmetry. Cosmologists and particle physicists look for additional violations of the CP-symmetry in the early universe that might account for the baryon asymmetry.

Both the problems of baryogenesis and cosmic inflation are very closely related to particle physics, and their resolution might come from high energy theory and experiment, rather than through observations of the universe.

Big Bang Theory

Big Bang nucleosynthesis is the theory of the formation of the elements in the early universe. It finished when the universe was about three minutes old and its temperature dropped below that at which nuclear fusion could occur. Big Bang nucleosynthesis had a brief period during which it could operate, so only the very lightest elements were produced. Starting from hydrogen ions (protons), it principally produced deuterium, helium-4, and lithium. Other elements were produced in only trace abundances. The basic theory of nucleosynthesis was developed in 1948 by George Gamow, Ralph Asher Alpher, and Robert Herman. It was used for many years as a probe of physics at the time of the Big Bang, as the theory of Big Bang nucleosynthesis connects the abundances of primordial light elements with the features of the early universe. Specifically, it can be used to test the equivalence principle, to probe dark matter, and test neutrino physics. Some cosmologists have proposed that Big Bang nucleosynthesis suggests there is a fourth "sterile" species of neutrino.

Standard Model of Big Bang Cosmology

The ΛCDM (Lambda cold dark matter) or Lambda-CDM model is a parametrization of the Big Bang cosmological model in which the universe contains a cosmological constant, denoted by Lambda (Greek Λ), associated with dark energy, and cold dark matter (abbreviated CDM). It is frequently referred to as the standard model of Big Bang cosmology.

Cosmic Microwave Background

The cosmic microwave background is radiation left over from decoupling after the epoch of recombination when neutral atoms first formed. At this point, radiation produced in the Big Bang stopped Thomson scattering from charged ions. The radiation, first observed in 1965 by Arno Penzias and Robert Woodrow Wilson, has a perfect thermal black-body spectrum. It has a temperature of 2.7 kelvins today and is isotropic to one part in 10^5. Cosmological perturbation theory, which describes the evolution of slight inhomogeneities in the early universe, has allowed cosmologists to precisely calculate the angular power spectrum of the radiation, and it has been measured by the recent satellite experiments (COBE and WMAP) and many ground and balloon-based experiments (such as Degree Angular Scale Interferometer, Cosmic Background Imager, and Boomerang). One of the goals of these efforts is to measure the basic parameters of the Lambda-CDM model with increasing accuracy, as well as to test the predictions of the Big Bang model and look for new physics. The recent measurements made by WMAP, for example, have placed limits on the neutrino masses.

Newer experiments, such as QUIET and the Atacama Cosmology Telescope, are trying to measure the polarization of the cosmic microwave background. These measurements are expected to provide further confirmation of the theory as well as information about cosmic inflation, and the so-called secondary anisotropies, such as the Sunyaev-Zel'dovich effect and Sachs-Wolfe effect, which are caused by interaction between galaxies and clusters with the cosmic microwave background.

On 17 March 2014, astronomers of the BICEP2 Collaboration announced the apparent detection of *B*-mode polarization of the CMB, considered to be evidence of primordial gravitational waves that are predicted by the theory of inflation to occur during the earliest phase of the Big Bang. However, later that year the Planck collaboration provided a more accurate measurement of cosmic dust, concluding that the B-mode signal from dust is the same strength as that reported from BICEP2. On January 30, 2015, a joint analysis of BICEP2 and Planck data was published and the European Space Agency announced that the signal can be entirely attributed to interstellar dust in the Milky Way.

Formation and Evolution of Large-scale Structure

Understanding the formation and evolution of the largest and earliest structures (i.e., quasars, galaxies, clusters and superclusters) is one of the largest efforts in cosmology. Cosmologists study a model of hierarchical structure formation in which structures form from the bottom up, with smaller objects forming first, while the largest objects, such as superclusters, are still assembling. One way to study structure in the universe is to survey the visible galaxies, in order to construct a three-dimensional picture of the galaxies in the universe and measure the matter power spectrum. This is the approach of the *Sloan Digital Sky Survey* and the 2dF Galaxy Redshift Survey.

Another tool for understanding structure formation is simulations, which cosmologists use to study the gravitational aggregation of matter in the universe, as it clusters into filaments, superclusters and voids. Most simulations contain only non-baryonic cold dark matter, which should suffice to understand the universe on the largest scales, as there is much more dark matter in the universe than visible, baryonic matter. More advanced simulations are starting to include baryons and study the formation of individual galaxies. Cosmologists study these simulations to see if they agree with the galaxy surveys, and to understand any discrepancy.

Other, complementary observations to measure the distribution of matter in the distant universe and to probe reionization include:

- The Lyman-alpha forest, which allows cosmologists to measure the distribution of neutral atomic hydrogen gas in the early universe, by measuring the absorption of light from distant quasars by the gas.
- The 21 centimeter absorption line of neutral atomic hydrogen also provides a sensitive test of cosmology.
- Weak lensing, the distortion of a distant image by gravitational lensing due to dark matter.

These will help cosmologists settle the question of when and how structure formed in the universe.

Dark Matter

Evidence from Big Bang nucleosynthesis, the cosmic microwave background and structure formation suggests that about 23% of the mass of the universe consists of non-baryonic dark matter, whereas only 4% consists of visible, baryonic matter. The gravitational effects of dark matter are well understood, as it behaves like a cold, non-radiative fluid that forms haloes around galaxies. Dark matter has never been detected in the laboratory, and the particle physics nature of dark matter remains completely unknown. Without observational constraints, there are a number of candidates, such as a stable supersymmetric particle, a weakly interacting massive particle, an axion, and a massive compact halo object. Alternatives to the dark matter hypothesis include a modification of gravity at small accelerations (MOND) or an effect from brane cosmology.

Dark Energy

If the universe is flat, there must be an additional component making up 73% (in addition to the 23% dark matter and 4% baryons) of the energy density of the universe. This is called dark energy. In order not to interfere with Big Bang nucleosynthesis and the cosmic microwave background, it must not cluster in haloes like baryons and dark matter. There is strong observational evidence for dark energy, as the total energy density of the universe is known through constraints on the flatness of the universe, but the amount of clustering matter is tightly measured, and is much less than this. The case for dark energy was strengthened in 1999, when measurements demonstrated that the expansion of the universe has begun to gradually accelerate.

Apart from its density and its clustering properties, nothing is known about dark energy. *Quantum field theory* predicts a cosmological constant (CC) much like dark energy, but 120 orders of magnitude larger than that observed. Steven Weinberg and a number of string theorists have invoked the 'weak anthropic principle': i.e. the reason that physicists observe a universe with such a small cosmological constant is that no physicists (or any life) could exist in a universe with a larger cosmological constant. Many cosmologists find this an unsatisfying explanation: perhaps because while the weak anthropic principle is self-evident (given that living observers exist, there must be at least one universe with a cosmological constant which allows for life to exist) it does not attempt to explain the context of that universe. For example, the weak anthropic principle alone does not distinguish between:

- Only one universe will ever exist and there is some underlying principle that constrains the CC to the value we observe.
- Only one universe will ever exist and although there is no underlying principle fixing the CC, we got lucky.
- Lots of universes exist (simultaneously or serially) with a range of CC values, and of course ours is one of the life-supporting ones.

Other possible explanations for dark energy include quintessence or a modification of gravity on the largest scales. The effect on cosmology of the dark energy that these models describe is given by the dark energy's equation of state, which varies depending upon the theory. The nature of dark energy is one of the most challenging problems in cosmology.

A better understanding of dark energy is likely to solve the problem of the ultimate fate of the universe. In the current cosmological epoch, the accelerated expansion due to dark energy is preventing structures larger than superclusters from forming. It is not known whether the acceleration will continue indefinitely, perhaps even increasing until a big rip, or whether it will eventually reverse.

Gravitational Waves

Gravitational waves are ripples in the curvature of spacetime that propagate as waves at the speed of light, generated in certain gravitational interactions that propagate outward from their source. Gravitational-wave astronomy is an emerging branch of observational astronomy which aims to use gravitational waves to collect observational data about sources of detectable gravitational waves such as binary star systems composed of white dwarfs, neutron stars, and black holes; and events such as supernovae, and the formation of the early universe shortly after the Big Bang.

In 2016, the LIGO Scientific Collaboration and Virgo Collaboration teams announced that they had made the first observation of gravitational waves, originating from a pair of merging black holes using the Advanced LIGO detectors. On June 15, 2016, a second detection of gravitational waves from coalescing black holes was announced. Besides LIGO, many other gravitational-wave observatories (detectors) are under construction.

Other Areas of Inquiry

Cosmologists also study:

- Whether primordial black holes were formed in our universe, and what happened to them.
- The GZK cutoff for high-energy cosmic rays, and whether it signals a failure of special relativity at high energies
- The equivalence principle, whether or not Einstein's general theory of relativity is the correct theory of gravitation, and if the fundamental laws of physics are the same everywhere in the universe.
- The increasing complexity of universal structures, an example being the progressively greater energy rate density.

Thermodynamics of the Universe

The thermodynamics of the universe is dictated by which form of energy dominates it - relativistic particles which are referred to as radiation, or non-relativistic particles which are referred to as matter. The former are particles whose rest mass is zero or negligible compared to their energy, and therefore move at the speed of light or very close to it; the latter are particles whose kinetic energy is much lower than their rest mass and therefore move much slower than the speed of light. The intermediate case is not treated well analytically.

Energy Density in the Expanding Universe

If the universe is expanding adiabatically then it will satisfy the first law of thermodynamics:

$$0 = dQ = dU + PdV$$

where Q is the total heat which is assumed to be constant, U is the internal energy of the matter and radiation in the universe, P is the pressure and V the volume.

One then finds an equation for the energy density $u \equiv U / V$, and so

$$du = d\left(\frac{U}{V}\right) = \frac{dU}{V} - U\frac{dV}{V^2} = -(p+u)\frac{dV}{V} = -3(p+u)\frac{da}{a}$$

where in the last equality we used the fact that the total volume of the universe is proportional to a^3, a being the scale factor of the universe.

In fact this equation can be directly obtained from the equations of motion governing the Friedmann-Lemaître-Robertson-Walker metric: by dividing the equation above with dt and identifying $\rho = u$ (the energy density), we get one of the FLRW equations of motions.

In the comoving coordinates, u is equal to the mass density ρ. For radiation, $p = u / 3$ whereas for matter $p << u$ and the pressure can be neglected. Thus we get,

For radiation $du = -4u\frac{da}{a}$ thus u is proportional to a^{-4}

For matter $du = -3u\frac{da}{a}$ thus u is proportional to a^{-3}

This can be understood as follows: For matter, the energy density is equal (in our approximation) to the rest mass density. This is inversely proportional to the volume, and is therefore proportional to a^{-3}. For radiation, the energy density depends on the temperature T as well, and is therefore proportional to Ta^{-3}. As the universe expands it cools down, so T depends on a as well. In fact, since the energy of a relativistic particle is inversely proportional to its wavelength, which is proportional to a, the energy density of the radiation must be proportional to a^{-4}.

From this discussion it is also obvious that the temperature of radiation is inversely proportional to the scale factor a.

Rate of Expansion of the Universe

Plugging this information to the Friedmann-Lemaître-Robertson-Walker equations of motion and neglecting both the cosmological constant Λ and the curvature parameter k, which is justified for the early universe ($a \ll 1$), one gets the following equation:

$$\dot{a}^2 \propto a^2 \rho$$

$\rho = u$ is the energy density, and one finds the following behavior:

- In a radiation-dominated universe: $a \propto t^{1/2}$
- In a matter-dominated universe: $a \propto t^{2/3}$

One can further show that the universe was radiation-dominated as long as the energy density was of the order of 10 eV to the fourth, or higher. Since the energy density keeps going down, this was no longer true when the universe was 70,000 years old, when it became matter dominant.

In the universe today, matter is mainly in forms of galaxies and dark matter, while the radiation is the cosmic microwave background radiation, the cosmic neutrino background (if the neutrino rest mass is high enough then the latter is formally matter), and finally, mostly in the form of dark energy.

Dark Energy and Cosmic Inflation

Dark energy is a hypothetical form of energy that permeates all of space and its negative pressure coincides with an acceleration in the expansion of the universe. Positive pressure coincides with a deceleration as does the gravity of energy and mass. There is no known cause and effect in fundamental physics, so it is not assumed the pressures or gravity "cause" a reduction or acceleration in the expansion of the universe, nor vice versa. For example, the energy in the gravitional field of the universe that coincides with its expansion is equal and opposite to the mass energy of the universe and it is not assumed (and the equations do not indicate) that the expansion created the positive mass energy and negative gravitational energy, nor vice versa.

According to the equation above,

$$\dot{u} = -3(p+u)\frac{\dot{a}}{a}$$

Thus the more negative the pressure is, the less the energy density reduces as the universe expands. In other words, Dark energy dilutes less than any other form of energy, and will therefore eventually dominate the universe, as all other energy densities gets diluted faster with the expansion of the universe.

In fact, if the dark energy is created by a cosmological constant or a constant scalar field, then its pressure is minus its energy density $p = -u$, and therefore its energy density remains constant (as is expected by definition).

Dark energy is usually assumed to be the Casimir energy of the vacuum, with possible contributions from the energy density of scalar fields which has a non-zero value at the vacuum. It may be that this field can decay at some time in the distant future, leading to a new vacuum state, different than the one we are living in. This is a phase transition, where the dark energy is reduced and huge amounts of energy in conventional forms (i.e. particles) are produced.

Such a series of events is in fact thought to have already occurred in the early universe, where first a cosmological constant much larger than the present one came to dominate the universe, bringing about cosmic inflation. At the end of this epoch, a phase transition occurred where the cosmological constant was reduced to its present value and huge amounts of energy where produced, from which all the radiation and matter of the early universe came about.

Particle Physics in Cosmology

Particle physics, which deals with the interactions of elementary particles at high energies, is an important component of cosmological models of the early universe, when the universe was dominated by radiation and its average energy density was very high. Because of this, pair production, scattering processes and decay of unstable particles are important in cosmology, and the interface between particle physics and cosmology is sometimes referred to as particle cosmology.

As a thumb rule, a scattering or a decay process is cosmologically important in a certain cosmological epoch if its relevant time scale is smaller or even to the time scale of the universe expansion, which is $\frac{1}{H}$ with H being the Hubble constant at that time. This is roughly equal to the age of the universe at that time.

For example, the pion has a lifetime of about 26 nanoseconds. This means that particle physics processes involving pions did not take place until roughly that much time passed since the start of the Big Bang.

Cosmological observations of phenomena such as the cosmic microwave background and the cosmic abundance of elements, together with the predictions of the Standard Model of particle physics, place constraints on the conditions of the early universe. The success of the Standard

Model at explaining these observations provides a confirmation of its validity outside of laboratory conditions. In addition, phenomena extrapolated from cosmological observations, such as dark matter and CP-violation, suggest a need for physics that goes beyond the Standard Model.

Friedmann–Lemaître–Robertson–Walker Metric

The Friedmann–Lemaître–Robertson–Walker (FLRW) metric is an exact solution of Einstein's field equations of general relativity; it describes a homogeneous, isotropic expanding or contracting universe that is path connected, but not necessarily simply connected. The general form of the metric follows from the geometric properties of homogeneity and isotropy; Einstein's field equations are only needed to derive the scale factor of the universe as a function of time. Depending on geographical or historical preferences, the set of the four scientists — Alexander Friedmann, Georges Lemaître, Howard P. Robertson and Arthur Geoffrey Walker are customarily grouped as Friedmann–Robertson–Walker (FRW) or Robertson–Walker (RW) or Friedmann–Lemaître (FL)). This model is sometimes called the *Standard Model* of modern cosmology, although such a description is also associated with the further developed Lambda-CDM model. The FLRW model was developed independently by the named authors in the 1920s and 1930s.

General Metric

The FLRW metric starts with the assumption of homogeneity and isotropy of space. It also assumes that the spatial component of the metric can be time-dependent. The generic metric which meets these conditions is,

$$-c^2 \mathrm{d}\tau^2 = -c^2 \mathrm{d}t^2 + a(t)^2 \mathrm{d}\Sigma^2$$

where Σ ranges over a 3-dimensional space of uniform curvature, that is, elliptical space, Euclidean space, or hyperbolic space. It is normally written as a function of three spatial coordinates, but there are several conventions for doing so, detailed below. $\mathrm{d}\Sigma$ does not depend on t — all of the time dependence is in the function $a(t)$, known as the "scale factor".

Reduced-circumference Polar Coordinates

In reduced-circumference polar coordinates the spatial metric has the form

$$\mathrm{d}\Sigma^2 = \frac{\mathrm{d}r^2}{1-kr^2} + r^2 \mathrm{d}\Omega^2, \quad \text{where } \mathrm{d}\Omega^2 = \mathrm{d}\theta^2 + \sin^2\theta \mathrm{d}\phi^2.$$

k is a constant representing the curvature of the space. There are two common unit conventions:

- k may be taken to have units of length^{-2}, in which case r has units of length and $a(t)$ is unitless. k is then the Gaussian curvature of the space at the time when $a(t) = 1$. r is sometimes called the reduced circumference because it is equal to the measured circumference of a circle (at that value of r), centered at the origin, divided by 2π (like the r of Schwarzschild coordinates). Where appropriate, $a(t)$ is often chosen to equal 1 in the present cosmological era, so that $d\Sigma$ measures comoving distance.

- Alternatively, k may be taken to belong to the set {−1,0,+1} (for negative, zero, and positive curvature respectively). Then r is unitless and $a(t)$ has units of length. When $k = \pm 1$, $a(t)$ is the radius of curvature of the space, and may also be written $R(t)$.

A disadvantage of reduced circumference coordinates is that they cover only half of the 3-sphere in the case of positive curvature—circumferences beyond that point begin to decrease, leading to degeneracy. (This is not a problem if space is elliptical, i.e. a 3-sphere with opposite points identified.)

Hyperspherical Coordinates

In *hyperspherical* or *curvature-normalized* coordinates the coordinate r is proportional to radial distance; this gives

$$d\Sigma^2 = dr^2 + S_k(r)^2 d\Omega^2$$

where $d\Omega$ is as before and

$$S_k(r) = \begin{cases} \sqrt{k}^{\,-1} \sin(r\sqrt{k}), & k > 0 \\ r, & k = 0 \\ \sqrt{|k|}^{\,-1} \sinh(r\sqrt{|k|}), & k < 0. \end{cases}$$

As before, there are two common unit conventions:

- k may be taken to have units of length^{-2}, in which case r has units of length and $a(t)$ is unitless. k is then the Gaussian curvature of the space at the time when $a(t) = 1$. Where appropriate, $a(t)$ is often chosen to equal 1 in the present cosmological era, so that $d\Sigma$ measures comoving distance.

- Alternatively, as before, k may be taken to belong to the set {−1,0,+1} (for negative, zero, and positive curvature respectively). Then r is unitless and $a(t)$ has units of length. When $k = \pm 1$, $a(t)$ is the radius of curvature of the space, and may also be written $R(t)$. Note that, when $k = +1$, r is essentially a third angle along with θ and φ. The letter χ may be used instead of r.

Though it is usually defined piecewise as above, S is an analytic function of both k and r. It can also be written as a power series,

$$S_k(r) = \sum_{n=0}^{\infty} \frac{(-1)^n k^n r^{2n+1}}{(2n+1)!} = r - \frac{kr^3}{6} + \frac{k^2 r^5}{120} - \cdots$$

or as

$$S_k(r) = r\,\mathrm{sinc}(r\sqrt{k})$$

where sinc is the unnormalized sinc function and $\sqrt{k}$ is one of the imaginary, zero or real square roots of k. These definitions are valid for all k.

Cartesian Coordinates

When k = 0 one may write simply

$$d\Sigma^2 = dx^2 + dy^2 + dz^2.$$

This can be extended to $k \neq 0$ by defining

$$x = r\cos\theta,$$

$$y = r\sin\theta\cos\phi \text{ , and}$$

$$z = r\sin\theta\sin\phi,$$

where r is one of the radial coordinates defined above, but this is rare.

Curvature

In flat (k=0) FRW space using Cartesian coordinates, the surviving components of the Ricci tensor are,

$$R_{tt} = -3\frac{\ddot{a}}{a}, \quad R_{xx} = R_{yy} = R_{zz} = c^{-2}(a\ddot{a} + 2\dot{a}^2)$$

and the Ricci scalar is,

$$R = 6c^{-2}\left(\frac{\ddot{a}(t)}{a(t)} + \frac{\dot{a}^2(t)}{a^2(t)}\right).$$

Spherical Coordinates

In more general FRW space using spherical coordinates (called "reduced-circumference polar coordinates" above), the surviving components of the Ricci tensor are,

$$R_{tt} = -3\frac{\ddot{a}}{a},$$

$$R_{rr} = \frac{c^{-2}(a(t)\ddot{a}(t) + 2\dot{a}^2(t)) + 2k}{1 - kr^2}$$

$$R_{\theta\theta} = r^2(c^{-2}(a(t)\ddot{a}(t) + 2\dot{a}^2(t)) + 2k)$$

$$R_{\phi\phi} = r^2(c^{-2}(a(t)\ddot{a}(t) + 2\dot{a}^2(t)) + 2k)\sin^2(\theta)$$

and the Ricci scalar is,

$$R = 6\left(\frac{\ddot{a}(t)}{c^2 a(t)} + \frac{\dot{a}^2(t)}{c^2 a^2(t)} + \frac{k}{a^2(t)}\right).$$

Solutions

Einstein's field equations are not used in deriving the general form for the metric: it follows from the geometric properties of homogeneity and isotropy. However, determining the time evolution of $a(t)$ does require Einstein's field equations together with a way of calculating the density, $\rho(t)$, such as a cosmological equation of state.

This metric has an analytic solution to Einstein's field equations $G_{\mu\nu} + \Lambda g_{\mu\nu} = \frac{8\pi G}{c^4} T_{\mu\nu}$ giving the Friedmann equations when the energy-momentum tensor is similarly assumed to be isotropic and homogeneous. The resulting equations are:

$$\left(\frac{\dot{a}}{a}\right)^2 + \frac{kc^2}{a^2} - \frac{\Lambda c^2}{3} = \frac{8\pi G}{3}\rho$$

$$2\frac{\ddot{a}}{a} + \left(\frac{\dot{a}}{a}\right)^2 + \frac{kc^2}{a^2} - \Lambda c^2 = -\frac{8\pi G}{c^2}p.$$

These equations are the basis of the standard big bang cosmological model including the current ΛCDM model. Because the FLRW model assumes homogeneity, some popular accounts mistakenly assert that the big bang model cannot account for the observed lumpiness of the universe. In a strictly FLRW model, there are no clusters of galaxies, stars or people, since these are objects much denser than a typical part of the universe. Nonetheless, the FLRW model is used as a first approximation for the evolution of the real, lumpy universe because it is simple to calculate, and models which calculate the lumpiness in the universe are added onto the FLRW models as extensions. Most cosmologists agree that the observable universe is well approximated by an *almost FLRW model*, i.e., a model which follows the FLRW metric apart from primordial density fluctuations. As of 2003, the theoretical implications of the various extensions to the FLRW model appear to be well understood, and the goal is to make these consistent with observations from COBE and WMAP.

If the spacetime is multiply connected, then each event in spacetime will be represented by more than one tuple of coordinates.

Interpretation

The pair of equations given above is equivalent to the following pair of equations,

$$\dot{\rho} = -3\frac{\dot{a}}{a}\left(\rho + \frac{p}{c^2}\right)$$

$$\frac{\ddot{a}}{a} = -\frac{4\pi G}{3}\left(\rho + \frac{3p}{c^2}\right) + \frac{\Lambda c^2}{3}$$

with k, the spatial curvature index, serving as a constant of integration for the first equation.

The first equation can be derived also from thermodynamical considerations and is equivalent to the first law of thermodynamics, assuming the expansion of the universe is an adiabatic process (which is implicitly assumed in the derivation of the Friedmann–Lemaître–Robertson–Walker metric).

The second equation states that both the energy density and the pressure cause the expansion rate of the universe $\dot{a}$ to decrease, i.e., both cause a deceleration in the expansion of the universe. This is a consequence of gravitation, with pressure playing a similar role to that of energy (or mass) density, according to the principles of general relativity. The cosmological constant, on the other hand, causes an acceleration in the expansion of the universe.

Cosmological Constant

The cosmological constant term can be omitted if we make the following replacements,

$$\rho \to \rho + \frac{\Lambda c^2}{8\pi G}$$

$$p \to p - \frac{\Lambda c^4}{8\pi G}.$$

Therefore, the cosmological constant can be interpreted as arising from a form of energy which has negative pressure, equal in magnitude to its (positive) energy density:

$$p = -\rho c^2.$$

Such form of energy—a generalization of the notion of a cosmological constant—is known as dark energy.

In fact, in order to get a term which causes an acceleration of the universe expansion, it is enough to have a scalar field which satisfies,

$$p < -\frac{\rho c^2}{3}.$$

Such a field is sometimes called quintessence.

Newtonian Interpretation

This is due to McCrea and Milne although sometimes incorrectly ascribed to Friedmann. The Friedmann equations are equivalent to this pair of equations:

$$-a^3\dot{\rho} = 3a^2\dot{a}\rho + \frac{3a^2 p\dot{a}}{c^2}$$

$$\frac{\dot{a}^2}{2} - \frac{G\frac{4\pi a^3}{3}\rho}{a} = -\frac{kc^2}{2}$$

The first equation says that the decrease in the mass contained in a fixed cube (whose side is momentarily a) is the amount which leaves through the sides due to the expansion of the universe plus the mass equivalent of the work done by pressure against the material being expelled. This is the conservation of mass-energy (first law of thermodynamics) contained within a part of the universe.

The second equation says that the kinetic energy (seen from the origin) of a particle of unit mass moving with the expansion plus its (negative) gravitational potential energy (relative to the mass contained in the sphere of matter closer to the origin) is equal to a constant related to the curvature of the universe. In other words, the energy (relative to the origin) of a co-moving particle in free-fall is conserved. General relativity merely adds a connection between the spatial curvature of the universe and the energy of such a particle: positive total energy implies negative curvature and negative total energy implies positive curvature.

The cosmological constant term is assumed to be treated as dark energy and thus merged into the density and pressure terms.

During the Planck epoch, one cannot neglect quantum effects. So they may cause a deviation from the Friedmann equations.

Name and History

The main results of the FLRW model were first derived by the Soviet mathematician Alexander Friedmann in 1922 and 1924. Although his work was published in the prestigious physics journal Zeitschrift für Physik, it remained relatively unnoticed by his contemporaries. Friedmann was in direct communication with Albert Einstein, who, on behalf of Zeitschrift für Physik, acted as the scientific referee of Friedmann's work. Eventually Einstein acknowledged the correctness of Friedmann's calculations, but failed to appreciate the physical significance of Friedmann's predictions.

Friedmann died in 1925. In 1927, Georges Lemaître, a Belgian priest, astronomer and periodic professor of physics at the Catholic University of Leuven, arrived independently at similar results as Friedmann had and published them in Annals of the Scientific Society of Brussels. In the face of the observational evidence for the expansion of the universe obtained by Edwin Hubble in the late 1920s, Lemaître's results were noticed in particular by Arthur Eddington, and in 1930–31 his paper was translated into English and published in the Monthly Notices of the Royal Astronomical Society.

Howard P. Robertson from the US and Arthur Geoffrey Walker from the UK explored the problem further during the 1930s. In 1935 Robertson and Walker rigorously proved that the FLRW metric is the only one on a spacetime that is spatially homogeneous and isotropic (as noted above, this is a geometric result and is not tied specifically to the equations of general relativity, which were always assumed by Friedmann and Lemaître).

Because the dynamics of the FLRW model were derived by Friedmann and Lemaître, the latter two names are often omitted by scientists outside the US. Conversely, US physicists often refer to it as simply "Robertson–Walker". The full four-name title is the most democratic and it is frequently used. Often the "Robertson–Walker" *metric*, so-called since they proved its generic properties, is distinguished from the dynamical "Friedmann-Lemaître" *models*, specific solutions for $a(t)$ which assume that the only contributions to stress-energy are cold matter ("dust"), radiation, and a cosmological constant.

Einstein's Radius of the Universe

Einstein's radius of the Universe is the radius of curvature of space of Einstein's universe, a long-abandoned static model that was supposed to represent our universe in idealized form. Putting,

$$\dot{a} = \ddot{a} = 0.$$

in the Friedmann equation, the radius of curvature of space of this universe (Einstein's radius) is,

$$R_E = c / \sqrt{4\pi G\rho},$$

where c is the speed of light, G is the Newtonian gravitational constant, and ρ is the density of space of this universe. The numerical value of Einstein's radius is of the order of 10^{10} light years.

Evidence

By combining the observation data from some experiments such as WMAP and Planck with theoretical results of Ehlers–Geren–Sachs theorem and its generalization, astrophysicists now agree that the universe is almost homogeneous and isotropic (when averaged over a very large scale) and thus nearly a FLRW spacetime.

Big Bang Nucleosynthesis

In physical cosmology, Big Bang nucleosynthesis (abbreviated BBN, also known as primordial nucleosynthesis, arch(a)eonucleosynthesis, archonucleosynthesis, protonucleosynthesis and pal(a)eonucleosynthesis) refers to the production of nuclei other than those of the lightest isotope of hydrogen (hydrogen-1, ^{1}H, having a single proton as a nucleus) during the early phases of the Universe. Primordial nucleosynthesis is believed by most cosmologists to have taken place in the interval from roughly 10 seconds to 20 minutes after the Big Bang, and is calculated to be responsible for the formation of most of the universe's helium as the isotope helium-4 (^{4}He), along with small amounts of the hydrogen isotope deuterium (^{2}H or D), the helium isotope helium-3 (^{3}He), and a very small amount of the lithium isotope lithium-7 (^{7}Li). In addition to these stable nuclei, two unstable or radioactive isotopes were also produced: the heavy hydrogen isotope tritium (^{3}H or T); and the beryllium isotope beryllium-7 (^{7}Be); but these unstable isotopes later decayed into ^{3}He and ^{7}Li, as above.

Essentially all of the elements that are heavier than lithium were created much later, by stellar nucleosynthesis in evolving and exploding stars.

Characteristics

There are several important characteristics of Big Bang nucleosynthesis (BBN):

- The initial conditions (neutron-proton ratio) were set in the first second after the Big Bang.
- The fusion of nuclei occurred between roughly 10 seconds to 20 minutes after the Big Bang; this corresponds to the temperature range when the universe was cool enough for deuterium to survive, but hot and dense enough for fusion reactions to occur at a significant rate.
- It was widespread, encompassing the entire observable universe.

The key parameter which allows one to calculate the effects of BBN is the baryon/photon number ratio, which is a small number of order 6×10^{-10}. This parameter corresponds to the baryon density and controls the rate at which nucleons collide and react; from this it is possible to calculate element abundances after nucleosynthesis ends Although the baryon per photon ratio is important in determining element abundances, the precise value makes little difference to the overall picture. Without major changes to the Big Bang theory itself, BBN will result in mass abundances of about 75% of hydrogen-1, about 25% helium-4, about 0.01% of deuterium and helium-3, trace amounts (on the order of 10^{-10}) of lithium, and negligible heavier elements. That the observed abundances in the universe are generally consistent with these abundance numbers is considered strong evidence for the Big Bang theory.

In this field, for historical reasons it is customary to quote the helium-4 fraction *by mass*, symbol Y, so that 25% helium-4 means that helium-4 atoms account for 25% of the mass, but less than 8% of the nuclei would be helium-4 nuclei. Other (trace) nuclei are usually expressed as number ratios to hydrogen.

Important Parameters

The creation of light elements during BBN was dependent on a number of parameters; among those was the neutron-proton ratio (calculable from Standard Model physics) and the baryon-photon ratio.

Neutron–proton Ratio

The neutron-proton ratio was set by Standard Model physics before the nucleosynthesis era, essentially within the first 1-second after the Big Bang. Neutrons can react with positrons or electron neutrinos to create protons and other products in one of the following reactions:

$$n + e^{+} \leftrightarrow \text{anti-}\nu_e + p$$

$$n + \nu_e \leftrightarrow p + e^{-}$$

At times much earlier than 1 sec, these reactions were fast and maintained the n/p ratio close to 1:1. As the temperature dropped, the equilibrium shifted in favour of protons due to their slightly lower mass, and the n/p ratio smoothly decreased. These reactions continued until the decreasing temperature and density caused the reactions to become too slow, which occurred at about T = 0.7 MeV (time around 1 second) and is called the freeze out temperature. At freeze out, the neutron-proton ratio was about 1/6. However, free neutrons are unstable with a mean life of 880 sec; some neutrons decayed in the next few minutes before fusing into any nucleus, so the ratio of total neutrons to protons after nucleosynthesis ends is about 1/7. Almost all neutrons that fused instead of decaying ended up combined into helium-4, due to the fact that helium-4 has the highest binding energy per nucleon among light elements. This predicts that about 8% of all atoms should be helium-4, leading to a mass fraction of helium-4 of about 25%, which is in line with observations. Small traces of deuterium and helium-3 remained as there was insufficient time and density for them to react and form helium-4.

Baryon–photon Ratio

The baryon–photon ratio, η, is the key parameter determining the abundances of light elements after nucleosynthesis ends. Baryons and light elements can fuse in the following main reactions:

$p + n \rightarrow {}^2H + \gamma$

$p + {}^2H \rightarrow {}^3He + \gamma$

${}^2H + {}^2H \rightarrow {}^3He + n$

${}^2H + {}^2H \rightarrow {}^3H + p$

${}^3He + {}^2H \rightarrow {}^4He + p$

${}^3H + {}^2H \rightarrow {}^4He + n$

along with some other low-probability reactions leading to ^7Li or ^7Be. (An important feature is that there are no stable nuclei with mass 5 or 8, which implies that reactions adding one baryon to ^4He, or fusing two ^4He, do not occur). Most fusion chains during BBN ultimately terminate in ^4He (helium-4), while "incomplete" reaction chains lead to small amounts of left-over ^2H or ^3He; the amount of these decreases with increasing baryon-photon ratio. That is, the larger the baryon-photon ratio the more reactions there will be and the more efficiently deuterium will be eventually transformed into helium-4. This result makes deuterium a very useful tool in measuring the baryon-to-photon ratio.

Sequence

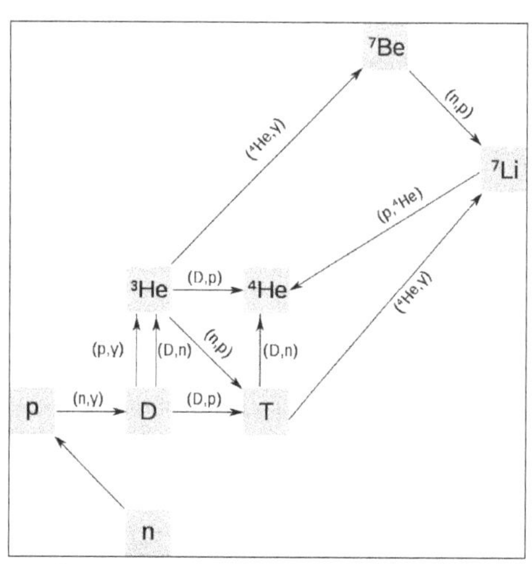

The main nuclear reaction chains for Big Bang nucleosynthesis.

Big Bang nucleosynthesis began roughly 10 seconds after the big bang, when the universe had cooled sufficiently to allow deuterium nuclei to survive disruption by high-energy photons. (Note that the neutron-proton freeze-out time was earlier). This time is essentially independent of dark matter content, since the universe was highly radiation dominated until much later, and this dominant component controls the temperature/time relation. At this time there were about six protons for every neutron, but a small fraction of the neutrons decay before fusing in the next few hundred seconds, so at the end of nucleosynthesis there are about seven protons to every neutron, and almost all the neutrons are in Helium-4 nuclei. The sequence of these reaction chains is shown on the image.

One feature of BBN is that the physical laws and constants that govern the behavior of matter at these energies are very well understood, and hence BBN lacks some of the speculative uncertainties that characterize earlier periods in the life of the universe. Another feature is that the process of nucleosynthesis is determined by conditions at the start of this phase of the life of the universe, and proceeds independently of what happened before.

As the universe expands, it cools. Free neutrons and protons are less stable than helium nuclei, and the protons and neutrons have a strong tendency to form helium-4. However, forming helium-4 requires the intermediate step of forming deuterium. Before nucleosynthesis began, the temperature was high enough for many photons to have energy greater than the binding energy of deuterium; therefore any deuterium that was formed was immediately destroyed (a situation known as the deuterium bottleneck). Hence, the formation of helium-4 is delayed until the universe became cool enough for deuterium to survive (at about T = 0.1 MeV); after which there was a sudden burst of element formation. However, very shortly thereafter, around twenty minutes after the Big Bang, the temperature and density became too low for any significant fusion to occur. At this point, the elemental abundances were nearly fixed, and the only changes were the result of the radioactive decay of the two major unstable products of BBN, tritium and beryllium-7.

History of Theory

The history of Big Bang nucleosynthesis began with the calculations of Ralph Alpher in the 1940s. Alpher published the Alpher–Bethe–Gamow paper that outlined the theory of light-element production in the early universe.

During the 1970s, there was a major puzzle in that the density of baryons as calculated by Big Bang nucleosynthesis was much less than the observed mass of the universe based on measurements of galaxy rotation curves and galaxy cluster dynamics. This puzzle was resolved in large part by postulating the existence of dark matter.

Heavy Elements

Big Bang nucleosynthesis produced no elements heavier than lithium, due to a bottleneck: the absence of a stable nucleus with 8 or 5 nucleons. This deficit of larger atoms also limited the amounts of lithium-7 produced during BBN. In stars, the bottleneck is passed by triple collisions of helium-4 nuclei, producing carbon (the triple-alpha process). However, this process is very slow and requires much higher densities, taking tens of thousands of years to convert a significant amount of helium to carbon in stars, and therefore it made a negligible contribution in the minutes following the Big Bang.

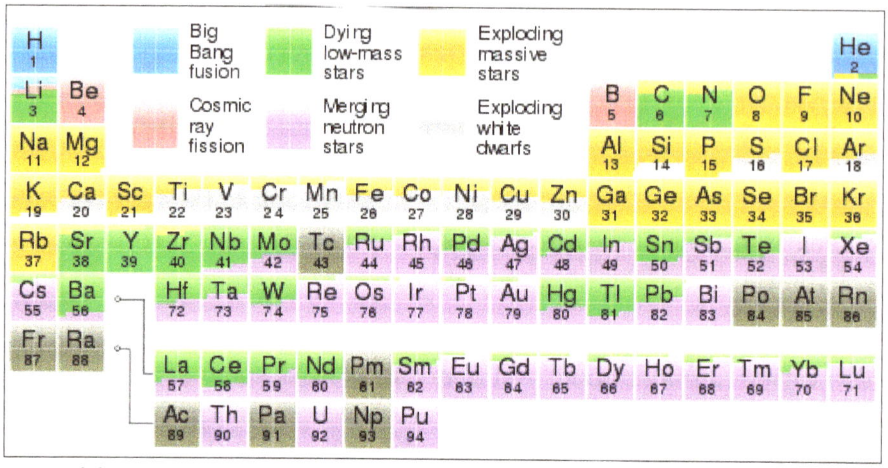

A version of the periodic table indicating the origins – including big bang nucleosynthesis – of the elements. All elements above 103 (lawrencium) are also manmade and are not included.

The predicted abundance of CNO isotopes produced in Big Bang nucleosynthesis is expected to be on the order of 10^{-15} that of H, making them essentially undetectable and negligible. Indeed, none of these primordial isotopes of the elements from lithium to oxygen have yet been detected, although those of beryllium and boron may be able to be detected in the future. So far, the only stable nuclides known experimentally to have been made before or during Big Bang nucleosynthesis are protium, deuterium, helium-3, helium-4, and lithium-7.

Helium-4

Big Bang nucleosynthesis predicts a primordial abundance of about 25% helium-4 by mass, irrespective of the initial conditions of the universe. As long as the universe was hot enough for protons and neutrons to transform into each other easily, their ratio, determined solely by their relative masses, was about 1 neutron to 7 protons (allowing for some decay of neutrons into protons). Once it was cool enough, the neutrons quickly bound with an equal number of protons to form first deuterium, then helium-4. Helium-4 is very stable and is nearly the end of this chain if it runs for only a short time, since helium neither decays nor combines easily to form heavier nuclei (since there are no stable nuclei with mass numbers of 5 or 8, helium does not combine easily with either protons, or with itself). Once temperatures are lowered, out of every 16 nucleons (2 neutrons and 14 protons), 4 of these (25% of the total particles and total mass) combine quickly into one helium-4 nucleus. This produces one helium for every 12 hydrogens, resulting in a universe that is a little over 8% helium by number of atoms, and 25% helium by mass.

One analogy is to think of helium-4 as ash, and the amount of ash that one forms when one completely burns a piece of wood is insensitive to how one burns it. The resort to the BBN theory of the helium-4 abundance is necessary as there is far more helium-4 in the universe than can be explained by stellar nucleosynthesis. In addition, it provides an important test for the Big Bang theory. If the observed helium abundance is significantly different from 25%, then this would pose a serious challenge to the theory. This would particularly be the case if the early helium-4 abundance was much smaller than 25% because it is hard to destroy helium-4. For a few years during the mid-1990s, observations suggested that this might be the case, causing astrophysicists to talk about a Big Bang nucleosynthetic crisis, but further observations were consistent with the Big Bang theory.

Deuterium

Deuterium is in some ways the opposite of helium-4 in that while helium-4 is very stable and very difficult to destroy, deuterium is only marginally stable and easy to destroy. The temperatures, time, and densities were sufficient to combine a substantial fraction of the deuterium nuclei to form helium-4 but insufficient to carry the process further using helium-4 in the next fusion step. BBN did not convert all of the deuterium in the universe to helium-4 due to the expansion that cooled the universe and reduced the density and so, cut that conversion short before it could proceed any further. One consequence of this is that unlike helium-4, the amount of deuterium is very sensitive to initial conditions. The denser the initial universe was, the more deuterium would be converted to helium-4 before time ran out, and the less deuterium would remain.

There are no known post-Big Bang processes which can produce significant amounts of deuterium. Hence observations about deuterium abundance suggest that the universe is not infinitely old, which is in accordance with the Big Bang theory.

During the 1970s, there were major efforts to find processes that could produce deuterium, but those revealed ways of producing isotopes other than deuterium. The problem was that while the concentration of deuterium in the universe is consistent with the Big Bang model as a whole, it is too high to be consistent with a model that presumes that most of the universe is composed of protons and neutrons. If one assumes that all of the universe consists of protons and neutrons, the density of the universe is such that much of the currently observed deuterium would have been burned into helium-4. The standard explanation now used for the abundance of deuterium is that the universe does not consist mostly of baryons, but that non-baryonic matter (also known as dark matter) makes up most of the mass of the universe. This explanation is also consistent with calculations that show that a universe made mostly of protons and neutrons would be far more *clumpy* than is observed.

It is very hard to come up with another process that would produce deuterium other than by nuclear fusion. Such a process would require that the temperature be hot enough to produce deuterium, but not hot enough to produce helium-4, and that this process should immediately cool to non-nuclear temperatures after no more than a few minutes. It would also be necessary for the deuterium to be swept away before it reoccurs.

Producing deuterium by fission is also difficult. The problem here again is that deuterium is very unlikely due to nuclear processes, and that collisions between atomic nuclei are likely to result either in the fusion of the nuclei, or in the release of free neutrons or alpha particles. During the 1970s, cosmic ray spallation was proposed as a source of deuterium. That theory failed to account for the abundance of deuterium, but led to explanations of the source of other light elements.

Measurements and Status of Theory

The theory of BBN gives a detailed mathematical description of the production of the light "elements" deuterium, helium-3, helium-4, and lithium-7. Specifically, the theory yields precise quantitative predictions for the mixture of these elements, that is, the primordial abundances at the end of the big-bang.

In order to test these predictions, it is necessary to reconstruct the primordial abundances as faithfully as possible, for instance by observing astronomical objects in which very little stellar nucleosynthesis has taken place (such as certain dwarf galaxies) or by observing objects that are very far away, and thus can be seen in a very early stage of their evolution (such as distant quasars).

As noted above, in the standard picture of BBN, all of the light element abundances depend on the amount of ordinary matter (baryons) relative to radiation (photons). Since the universe is presumed to be homogeneous, it has one unique value of the baryon-to-photon ratio. For a long time, this meant that to test BBN theory against observations one had to ask: can *all* of the light element observations be explained with a *single value* of the baryon-to-photon ratio? Or more precisely, allowing for the finite precision of both the predictions and the observations, one asks: is there some *range* of baryon-to-photon values which can account for all of the observations?

More recently, the question has changed: Precision observations of the cosmic microwave background radiation with the Wilkinson Microwave Anisotropy Probe (WMAP) and Planck give an independent value for the baryon-to-photon ratio. Using this value, are the BBN predictions for the abundances of light elements in agreement with the observations?

The present measurement of helium-4 indicates good agreement, and yet better agreement for helium-3. But for lithium-7, there is a significant discrepancy between BBN and WMAP/Planck, and the abundance derived from Population II stars. The discrepancy is a factor of 2.4–4.3 below the theoretically predicted value and is considered a problem for the original models, that have resulted in revised calculations of the standard BBN based on new nuclear data, and to various reevaluation proposals for primordial proton-proton nuclear reactions, especially the abundances of $^{7}Be + n \rightarrow {}^{7}Li + p$, versus $^{7}Be + {}^{2}H \rightarrow {}^{8}Be + p$.

Non-standard Scenarios

In addition to the standard BBN scenario there are numerous non-standard BBN scenarios. These should not be confused with non-standard cosmology: a non-standard BBN scenario assumes that the Big Bang occurred, but inserts additional physics in order to see how this affects elemental abundances. These pieces of additional physics include relaxing or removing the assumption of homogeneity, or inserting new particles such as massive neutrinos.

There have been, and continue to be, various reasons for researching non-standard BBN. The first, which is largely of historical interest, is to resolve inconsistencies between BBN predictions and observations. This has proved to be of limited usefulness in that the inconsistencies were resolved by better observations, and in most cases trying to change BBN resulted in abundances that were more inconsistent with observations rather than less. The second reason for researching non-standard BBN, and largely the focus of non-standard BBN in the early 21st century, is to use BBN to place limits on unknown or speculative physics. For example, standard BBN assumes that no exotic hypothetical particles were involved in BBN. One can insert a hypothetical particle (such as a massive neutrino) and see what has to happen before BBN predicts abundances that are very different from observations. This has been done to put limits on the mass of a stable tau neutrino.

Absolute Time and Space

Absolute space and time is a concept in physics and philosophy about the properties of the universe. In physics, absolute space and time may be a preferred frame.

Before Newton

A version of the concept of absolute space (in the sense of a preferred frame) can be seen in Aristotelian physics. Robert S. Westman writes that "whiff" of absolute space can be observed in Copernicus De revolutionibus orbium coelestium, where he exploits the concept of immobile sphere of stars.

Newton

Originally introduced by Sir Isaac Newton in *Philosophiæ Naturalis Principia Mathematica*, the concepts of absolute time and space provided a theoretical foundation that facilitated Newtonian mechanics. According to Newton, absolute time and space respectively are independent aspects of objective reality:

Absolute, true and mathematical time, of itself, and from its own nature flows equably without regard to anything external, and by another name is called duration: relative, apparent and common time, is some sensible and external (whether accurate or unequable) measure of duration by the means of motion, which is commonly used instead of true time ...

According to Newton, absolute time exists independently of any perceiver and progresses at a consistent pace throughout the universe. Unlike relative time, Newton believed absolute time was imperceptible and could only be understood mathematically. According to Newton, humans are only capable of perceiving relative time, which is a measurement of perceivable objects in motion (like the Moon or Sun). From these movements, we infer the passage of time.

Absolute space, in its own nature, without regard to anything external, remains always similar and immovable. Relative space is some movable dimension or measure of the absolute spaces; which our senses determine by its position to bodies: and which is vulgarly taken for immovable space ... Absolute motion is the translation of a body from one absolute place into another: and relative motion, the translation from one relative place into another ...

—Isaac Newton

These notions imply that absolute space and time do not depend upon physical events, but are a backdrop or stage setting within which physical phenomena occur. Thus, every object has an absolute state of motion relative to absolute space, so that an object must be either in a state of absolute rest, or moving at some absolute speed. To support his views, Newton provided some empirical examples: according to Newton, a solitary rotating sphere can be inferred to rotate about its axis relative to absolute space by observing the bulging of its equator, and a solitary pair of spheres tied by a rope can be inferred to be in absolute rotation about their center of gravity (barycenter) by observing the tension in the rope.

Absolute time and space continue to be used in classical mechanics, but modern formulations by authors such as Walter Noll and Clifford Truesdell go beyond the linear algebra of elastic moduli to use topology and functional analysis for non-linear field theories.

Differing Views

Historically, there have been differing views on the concept of absolute space and time. Gottfried Leibniz was of the opinion that space made no sense except as the relative location of bodies, and time made no sense except as the relative movement of bodies. George Berkeley suggested that, lacking any point of reference, a sphere in an otherwise empty universe could not be conceived to rotate, and a pair of spheres could be conceived to rotate relative to one another, but not to rotate about their center of gravity, an example later raised by Albert Einstein in his development of general relativity.

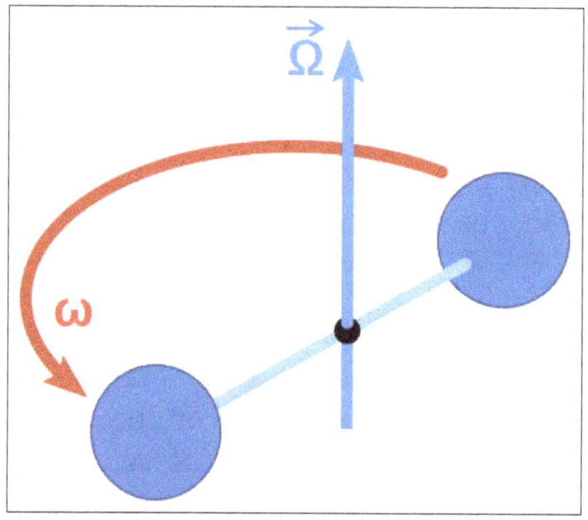

Two spheres orbiting around an axis. The spheres are distant enough for their effects on each other to be ignored, and they are held together by a rope. The rope is under tension if the bodies are rotating relative to absolute space according to Newton, or because they rotate relative to the universe itself according to Mach, or because they rotate relative to local geodesics according to general relativity.

A more recent form of these objections was made by Ernst Mach. Mach's principle proposes that mechanics is entirely about relative motion of bodies and, in particular, mass is an expression of such relative motion. So, for example, a single particle in a universe with no other bodies would have zero mass. According to Mach, Newton's examples simply illustrate relative rotation of spheres and the bulk of the universe.

> When, accordingly, we say that a body preserves unchanged its direction and velocity *in space*, our assertion is nothing more or less than an abbreviated reference to *the entire universe*.
>
> —Ernst Mach; as quoted by Ciufolini and Wheeler: *Gravitation and Inertia*, p. 387

These views opposing absolute space and time may be seen from a modern stance as an attempt to introduce operational definitions for space and time, a perspective made explicit in the special theory of relativity.

Even within the context of Newtonian mechanics, the modern view is that absolute space is unnecessary. Instead, the notion of inertial frame of reference has taken precedence, that is, a preferred *set* of frames of reference that move uniformly with respect to one another. The laws of physics transform from one inertial frame to another according to Galilean relativity, leading to the following objections to absolute space, as outlined by Milutin Blagojević:

- The existence of absolute space contradicts the internal logic of classical mechanics since, according to Galilean principle of relativity, none of the inertial frames can be singled out.
- Absolute space does not explain inertial forces since they are related to acceleration with respect to any one of the inertial frames.

- Absolute space acts on physical objects by inducing their resistance to acceleration but it cannot be acted upon.

Newton himself recognized the role of inertial frames.

The motions of bodies included in a given space are the same among themselves, whether that space is at rest or moves uniformly forward in a straight line.

As a practical matter, inertial frames often are taken as frames moving uniformly with respect to the fixed stars.

Special Relativity

The concepts of space and time were separate in physical theory prior to the advent of special relativity theory, which connected the two and showed both to be dependent upon the reference frame's motion. In Einstein's theories, the ideas of absolute time and space were superseded by the notion of spacetime in special relativity, and curved spacetime in general relativity.

Absolute simultaneity refers to the concurrence of events in time at different locations in space in a manner agreed upon in all frames of reference. The theory of relativity does not have a concept of absolute time because there is a relativity of simultaneity. An event that is simultaneous with another event in one frame of reference may be in the past or future of that event in a different frame of reference, which negates absolute simultaneity.

Einstein

Quoted below from his later papers, Einstein identified the term aether with "properties of space", a terminology that is not widely used. Einstein stated that in general relativity the "aether" is not absolute anymore, as the geodesic and therefore the structure of spacetime depends on the presence of matter.

To deny the ether is ultimately to assume that empty space has no physical qualities whatever. The fundamental facts of mechanics do not harmonize with this view. For the mechanical behaviour of a corporeal system hovering freely in empty space depends not only on relative positions (distances) and relative velocities, but also on its state of rotation, which physically may be taken as a characteristic not appertaining to the system in itself. In order to be able to look upon the rotation of the system, at least formally, as something real, Newton objectivises space. Since he classes his absolute space together with real things, for him rotation relative to an absolute space is also something real. Newton might no less well have called his absolute space "Ether"; what is essential is merely that besides observable objects, another thing, which is not perceptible, must be looked upon as real, to enable acceleration or rotation to be looked upon as something real.

—Albert Einstein, Ether and the Theory of Relativity (1920)

Because it was no longer possible to speak, in any absolute sense, of simultaneous states at different locations in the aether, the aether became, as it were, four-dimensional, since there was no objective way of ordering its states by time alone. According to special relativity too, the aether was absolute, since its influence on inertia and the propagation of light was thought of as being itself independent of physical influence....The theory of relativity resolved this problem by establishing the behaviour of the electrically neutral point-mass by the law of the geodetic line, according to which inertial and gravitational effects are no longer considered as separate. In doing so, it attached characteristics to the aether which vary from point to point, determining the metric and the dynamic behaviour of material points, and determined, in their turn, by physical factors, namely the distribution of mass/energy. Thus the aether of general relativity differs from those of classical mechanics and special relativity in that it is not 'absolute' but determined, in its locally variable characteristics, by ponderable matter.

—Albert Einstein, Über den Äther (1924)

General Relativity

Special relativity eliminates absolute time (although Gödel and others suspect absolute time may be valid for some forms of general relativity) and general relativity further reduces the physical scope of absolute space and time through the concept of geodesics. There appears to be absolute space in relation to the distant stars because the local geodesics eventually channel information from these stars, but it is not necessary to invoke absolute space with respect to any system's physics.

De Revolutionibus Orbium Coelestium

De revolutionibus orbium coelestium (*On the Revolutions of the Heavenly Spheres*) is the seminal work on the heliocentric theory of the Renaissance astronomer Nicolaus Copernicus (1473–1543). The book, first printed in 1543 in Nuremberg, Holy Roman Empire, offered an alternative model of the universe to Ptolemy's geocentric system, which had been widely accepted since ancient times.

History

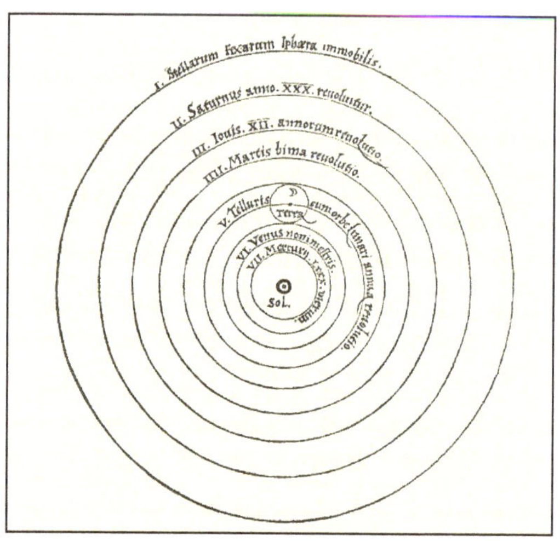

Heliocentric model of the solar system in Copernicus' manuscript.

Copernicus initially outlined his system in a short, untitled, anonymous manuscript that he distributed to several friends, referred to as the *Commentariolus*. A physician's library list dating to 1514 includes a manuscript whose description matches the *Commentariolus*, so Copernicus must have begun work on his new system by that time. Most historians believe that he wrote the *Commentariolus* after his return from Italy, possibly only after 1510. At this time, Copernicus anticipated that he could reconcile the motion of the Earth with the perceived motions of the planets easily, with fewer motions than were necessary in the *Alfonsine Tables*, the version of the Ptolemaic system current at the time. In particular, the heliocentric Copernican model made use of the Urdi Lemma developed in the 13th century by Mu'ayyad al-Din al-'Urdi, the first of the Maragha astronomers to develop a non-Ptolemaic model of planetary motion.

Observations of Mercury by Bernhard Walther (1430–1504) of Nuremberg, a pupil of Regiomontanus, were made available to Copernicus by Johannes Schöner, 45 observations in total, 14 of them with longitude and latitude. Copernicus used three of them in *De revolutionibus*, giving only longitudes, and erroneously attributing them to Schöner. Copernicus' values differed slightly from the ones published by Schöner in 1544 in *Observationes XXX annorum a I. Regiomontano et B. Walthero Norimbergae habitae, [4°, Norimb. 1544].*

A manuscript of *De revolutionibus* in Copernicus' own hand has survived. After his death, it was given to his pupil, Rheticus, who for publication had only been given a copy without annotations. Via Heidelberg, it ended up in Prague, where it was rediscovered and studied in the 19th century. Close examination of the manuscript, including the different types of paper used, helped scholars construct an approximate timetable for its composition. Apparently Copernicus began by making a few astronomical observations to provide new data to perfect his models. He may have begun writing the book while still engaged in observations. By the 1530s a substantial part of the book was complete, but Copernicus hesitated to publish.

In 1539 Georg Joachim Rheticus, a young mathematician from Wittenberg, arrived in Frauenburg (Frombork) to study with him. Rheticus read Copernicus' manuscript and immediately wrote a non-technical summary of its main theories in the form of an open letter addressed to Schöner, his astrology teacher in Nürnberg; he published this letter as the *Narratio Prima* in Danzig in 1540. Rheticus' friend and mentor Achilles Gasser published a second edition of the *Narratio* in Basel in 1541. Due to its friendly reception, Copernicus finally agreed to publication of more of his main work—in 1542, a treatise on trigonometry, which was taken from the second book of the still unpublished *De revolutionibus*. Rheticus published it in Copernicus' name.

Under strong pressure from Rheticus, and having seen that the first general reception of his work had not been unfavorable, Copernicus finally agreed to give the book to his close friend, Bishop Tiedemann Giese, to be delivered to Rheticus in Wittenberg for printing by Johannes Petreius at Nürnberg (Nuremberg). It was published just before Copernicus' death, in 1543.

The book is dedicated to Pope Paul III in a preface that argues that mathematics, not physics, should be the basis for understanding and accepting his new theory.

De revolutionibus is divided into six "books" (sections or parts), following closely the layout of Ptolemy's *Almagest* which it updated and replaced:

- Book I chapters 1–11 are a general vision of the heliocentric theory, and a summarized exposition of his cosmology. The world (heavens) is spherical, as is the earth, and the land and water make a single globe. The celestial bodies, including the earth, have regular circular and everlasting movements. The earth rotates on its axis and around the sun. Answers to why the ancients thought the earth was central. The order of the planets around the sun and their periodicity. Chapters 12-14 give theorems for chord geometry as well as a table of chords.

- Book II describes the principles of spherical astronomy as a basis for the arguments developed in the following books and gives a comprehensive catalogue of the fixed stars.
- Book III describes his work on the precession of the equinoxes and treats the apparent movements of the Sun and related phenomena.
- Book IV is a similar description of the Moon and its orbital movements.
- Book V explains how to calculate the positions of the wandering stars based on the heliocentric model and gives tables for the five planets.
- Book VI deals with the digression in latitude from the ecliptic of the five planets.

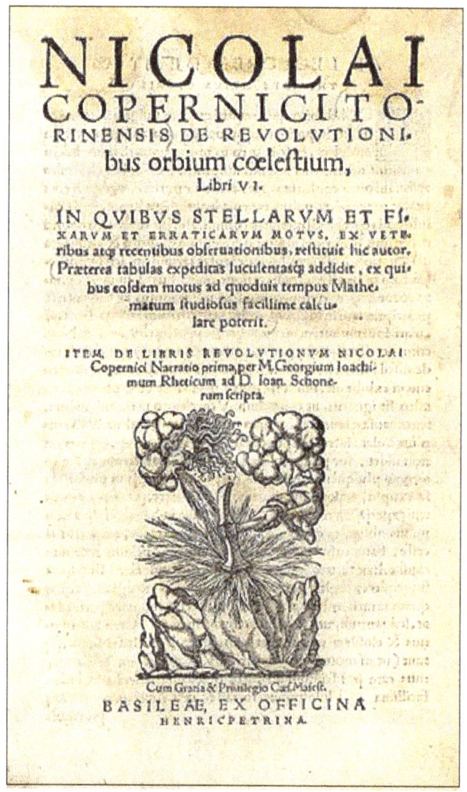
NICOLAI
COPERNICI TO-
RINENSIS DE REVOLVTIONI-
bus orbium cœlestium,
Libri VI.
IN QVIBVS STELLARVM ET FI-
XARVM ET ERRATICARVM MOTVS, EX VETE-
ribus atq; recentibus obseruationibus, restituit hic autor.
Præterea tabulas expeditas luculentasq; addidit, ex qui-
bus eosdem motus ad quoduis tempus Mathe-
matum studiosus facillime calcu-
lare poterit.
ITEM, DE LIBRIS REVOLVTIONVM NICOLAI
Copernici Narratio prima, per M. Georgium Ioachi-
mum Rheticum ad D. Ioan. Schone-
rum scripta.
Cum Gratia & Priuilegio Cæs. Maiest.
BASILEAE, EX OFFICINA
HENRICPETRINA.

Title page, 2nd edition, Basel, *Officina Henricpetrina*, 1566.

Copernicus argued that the universe comprised eight spheres. The outermost consisted of motionless, fixed stars, with the Sun motionless at the center. The known planets revolved about the Sun, each in its own sphere, in the order: Mercury, Venus, Earth, Mars, Jupiter, Saturn. The Moon, however, revolved in its sphere around the Earth. What appeared to be the daily revolution of the Sun and fixed stars around the Earth was actually the Earth's daily rotation on its own axis.

Copernicus adhered to one of the standard beliefs of his time, namely that the motions of celestial bodies must be composed of uniform circular motions. For this reason, he was unable to account for the observed apparent motion of the planets without retaining a complex system of epicycles similar to those of the Ptolemaic system. Despite Copernicus' adherence to this aspect of ancient astronomy, his radical shift from a geocentric to a heliocentric cosmology was a serious blow to Aristotle's science—and helped usher in the Scientific Revolution.

Ad Lectorem

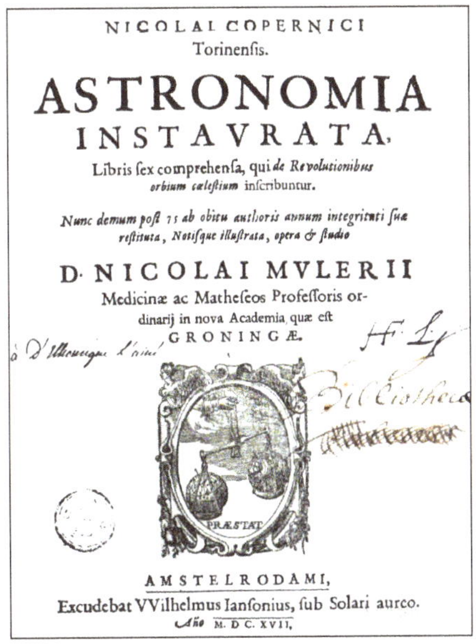
NICOLAI COPERNICI
Torinensis.
ASTRONOMIA
INSTAVRATA,
Libris sex comprehensa, qui de Revolutionibus orbium cælestium inscribuntur.
Nunc demum post 75 ab obitu authoris annum integritati suæ restituta, Notisque illustrata, opera & studio
D. NICOLAI MVLERII
Medicinæ ac Matheseos Professoris ordinarij in nova Academia quæ est
GRONINGÆ.

AMSTELRODAMI,
Excudebat VVilhelmus Ianßonius, sub Solari aureo.
Anno M. DC. XVII.

Title page, 3rd ed., Amsterdam, Nicolaus Mulerius, publisher, 1617.

Rheticus left Nürnberg to take up his post as professor in Leipzig. The Lutheran preacher Andreas Osiander had taken over the task of supervising the printing and publication. In an effort to reduce the controversial impact of the book Osiander added his own unsigned letter *Ad lectorem de hypothesibus huius operis* (*To the reader concerning the hypotheses of this work*) printed in front of Copernicus' preface which was a dedicatory letter to Pope Paul III and which kept the title "Praefatio authoris" (to acknowledge that the unsigned letter was not by the book's author).

Osiander's letter stated that Copernicus' system was mathematics intended to aid computation and not an attempt to declare literal truth:

it is the duty of an astronomer to compose the history of the celestial motions through careful and expert study. Then he must conceive and devise the causes of these motions or hypotheses about them. Since he cannot in any way attain to the true causes, he will adopt whatever suppositions enable the motions to be computed correctly ... The present author has performed both these duties excellently. For these hypotheses need not be true nor even probable. On the contrary, if they provide a calculus consistent with the observations, that alone is enough ... For this art, it is quite clear, is completely and absolutely ignorant of the causes of the apparent [movement of the heavens]. And if any causes are devised by the imagination, as indeed very many are, they are not put forward to convince anyone that they are true, but merely to provide a reliable basis for computation. However, since different hypotheses are sometimes offered for one and the same ... the astronomer will take as his first choice that hypothesis which is the easiest to grasp. The philosopher will perhaps rather seek the semblance of the truth. But neither of them will understand or state anything certain, unless it has been divinely revealed to him ... Let no one expect anything certain from astronomy, which cannot furnish it, lest he accept as the truth ideas conceived for another purpose, and depart this study a greater fool than when he entered.

As even Osiander's defenders point out, the *Ad lectorem* "expresses views on the aim and nature of scientific theories at variance with Copernicus' claims for his own theory".

Many view Osiander's letter as a betrayal of science and Copernicus, and an attempt to pass his own thoughts off as those of the book's author. An example of this type of claim can be seen in the *Catholic Encyclopedia,* which states "Fortunately for him [the dying Copernicus], he could not see what Osiander had done. This reformer, knowing the attitude of Luther and Melanchthon against the heliocentric system ... without adding his own name, replaced the preface of Copernicus by another strongly contrasting in spirit with that of Copernicus."

While Osiander's motives behind the letter have been questioned by many, he has been defended by historian Bruce Wrightsman, who points out he was not an enemy of science. Osiander had many scientific connections including "Johannes Schoner, Rheticus's teacher, whom Osiander recommended for his post at the Nurnberg Gymnasium; Peter Apian of Ingolstadt University; Hieronymous Schreiber...Joachim Camerarius...Erasmus Reinhold...Joachim Rheticus...and finally, Hieronymous Cardan."

The historian Wrightsman put forward that Osiander did not sign the letter because he "was such a notorious [Protestant] reformer whose name was well-known and infamous among Catholics", so that signing would have likely caused negative scrutiny of the work of Copernicus (a loyal Catholic canon and scholar). Copernicus himself had communicated to Osiander his "own fears that his work would be scrutinized and criticized by the 'peripatetics and theologians'," and he had already been in trouble with his bishop, Johannes Dantiscus, on account of his former relationship with his mistress and friendship with Dantiscus's enemy and suspected heretic, Alexander Scultetus. It was also possible that Protestant Nurnberg could fall to the forces of the Holy Roman Emperor and since "the books of hostile theologians could be burned...why not scientific works with the names of hated theologians affixed to them?" Wrightsman also holds that this is why Copernicus did not mention his top student, Rheticus (a Lutheran) in the book's dedication to the Pope.

Osiander's interest in astronomy was theological, hoping for "improving the chronology of historical events and thus providing more accurate apocalyptic interpretations of the Bible... [he shared in] the general awareness that the calendar was not in agreement with astronomical movement and therefore, needed to be corrected by devising better models on which to base calculations." In an era before the telescope, Osiander (like most of the era's mathematical astronomers) attempted to bridge the "fundamental incompatibility between Ptolemaic astronomy and Aristotlian physics, and the need to preserve both", by taking an 'instrumentalist' position. Only the handful of "Philosophical purists like the Averroists... demanded physical consistency and thus sought for realist models."

Copernicus was hampered by his insistence on preserving the idea that celestial bodies had to travel in perfect spheres – he "was still attached to classical ideas of circular motion around deferents and epicycles, and spheres." This was particularly troubling concerning the Earth because he "attached the Earth's axis rigidly to a Sun-centered sphere. The unfortunate consequence was that the terrestrial rotation axis then maintained the same inclination with respect to the Sun as the sphere turned, eliminating the seasons." To explain the seasons, he had to propose a third motion, "an annual contrary conical sweep of the terrestrial axis". It was not until the Great Comet of 1577, which moved as if there were no spheres to crash through, did the idea come under question. In 1609, Kepler fixed Copernicus' theory by stating that the planets orbit the sun not in circles, but ellipses. Only after Kepler's refinement of Copernicus' theory was the need for deferents and epicycles abolished.

In his work, Copernicus "used conventional, hypothetical devices like epicycles...as all astronomers had done since antiquity. ...hypothetical constructs solely designed to 'save the phenomena' and aid computation". Ptolemy's theory contained a hypothesis about the epicycle of Venus that was viewed as absurd if seen as anything other than a geometrical device (its brightness and distance should have varied greatly, but they don't). "In spite of this defect in Ptolemy's theory, Copernicus' hypothesis predicts approximately the same variations." Because of the use of similar terms and similar deficiencies, Osiander could see "little technical or physical truth-gain" between one system and the other. It was this attitude towards technical astronomy that had allowed it to "function since antiquity, despite its inconsistencies with the principles of physics and the philosophical objections of Averroists."

Writing *Ad lectorem*, Osiander was influenced by Pico della Mirandola's idea that humanity "orders [an intellectual] cosmos out of the chaos of opinions." From Pico's writings, Osiander "learned to extract and synthesize insights from many sources without becoming the slavish follower of any of them." The effect of Pico on Osiander was tempered by the influence of Nicholas of Cusa's and his idea of *coincidentia oppositorum*. Rather than having Pico's focus on human effort, Osiander followed Cusa's idea that understanding the Universe and its Creator only came from divine inspiration rather than intellectual organization. From these influences, Osiander held that in the area of philosophical speculation and scientific hypothesis there are "no heretics of the intellect", but when one gets past speculation into truth-claims the Bible is the ultimate measure. By holding Copernicianism was mathematical speculation, Osiander held that it would be silly to hold it up against the accounts of the Bible.

Pico's influence on Osiander did not escape Rheticus, who reacted strongly against the *Ad lectorem*. As historian Robert S. Westman puts it, "The more profound source of Rheticus's ire however, was Osiander's view of astronomy as a disciple fundamentally incapable of knowing anything with certainty. For Rheticus, this extreme position surely must have resonated uncomfortably with Pico della Mirandola's attack on the foundations of divinatory astrology."

In his *Disputations*, Pico had made a devastating attack on astrology. Because those who were making astrological predictions relied on astronomers to tell them where the planets were, they also became a target. Pico held that since astronomers who calculate planetary positions could not agree among themselves, how were they to be held as reliable? While Pico could bring into concordance writers like Aristotle, Plato, Plotinus, Averroes, Avicenna, and Aquinas, the lack of consensus he saw in astronomy was a proof to him of its fallibility alongside astrology. Pico pointed out that the astronomers' instruments were imprecise and any imperfection of even a degree made them worthless for astrology, people should not trust astrologists because they should not trust the numbers from astronomers. Pico pointed out that astronomers couldn't even tell where the sun appeared in the order of the planets as they orbited the earth (some put it close to the moon, others among the planets). How, Pico asked, could astrologists possibly claim they could read what was going on when the astronomers they relied on could offer no precision on even basic questions?

As Westman points out, to Rheticus "it would seem that Osiander now offered new grounds for endorsing Pico's conclusions: not merely was the disagreement among astronomers grounds for mistrusting the sort of knowledge that they produced, but now Osiander proclaimed that astronomers might construct a world deduced from (possibly) false premises. Thus the conflict between Piconian skepticism and secure principles for the science of the stars was built right into the complex dedicatory apparatus of *De Revolutionibus* itself." According to the notes of Michael Maestlin,

"Rheticus...became embroiled in a very bitter wrangle with the printer [over the Ad lectorem]. Rheticus...suspected Osiander had prefaced the work; if he knew this for certain, he declared, he would rough up the fellow so violently that in future he would mind his own business."

Objecting to the *Ad lectorem*, Tiedemann Giese urged the Nuremberg city council to issue a correction, but this was not done, and the matter was forgotten. Jan Broscius, a supporter of Copernicus, also despaired of the *Ad lectorem*, writing "Ptolemy's hypothesis is the earth rests. Copernicus' hypothesis is that the earth is in motion. Can either, therefore, be true? ... Indeed, Osiander deceives much with that preface of his ... Hence, someone may well ask: How is one to know which hypothesis is truer, the Ptolemaic or the Copernican?"

Petreius had sent a copy to Hieronymus Schreiber, an astronomer from Nürnberg who had substituted for Rheticus as professor of mathematics in Wittenberg while Rheticus was in Nürnberg supervising the printing. Schreiber, who died in 1547, left in his copy of the book a note about Osiander's authorship. Via Michael Mästlin, this copy came to Johannes Kepler, who discovered what Osiander had done and methodically demonstrated that Osiander had indeed added the foreword. The most knowledgeable astronomers of the time had realized that the foreword was Osiander's doing.

Owen Gingerich gives a slightly different version: Kepler knew of Osiander's authorship since he had read about it in one of Schreiber's annotations in his copy of *De Revolutionibus*; Maestlin learned of the fact from Kepler. Indeed, Maestlin perused Kepler's book, up to the point of leaving a few annotations in it. However, Maestlin already suspected Osiander, because he had bought his *De revolutionibus* from the widow of Philipp Apian; examining his books, he had found a note attributing the introduction to Osiander.

Johannes Praetorius (1537–1616), who learned of Osiander's authorship from Rheticus during a visit to him in Kraków, wrote Osiander's name in the margin of the foreword in his copy of *De revolutionibus*.

All three early editions of *De revolutionibus* included Osiander's foreword.

Reception

Even before the 1543 publication of *De revolutionibus*, rumors circulated about its central theses. Martin Luther is quoted as saying in 1539:

People gave ear to an upstart astrologer who strove to show that the earth revolves, not the heavens or the firmament, the sun and the moon ... This fool wishes to reverse the entire science of astronomy; but sacred Scripture tells us [Joshua 10:13] that Joshua commanded the sun to stand still, and not the earth.

When the book was finally published, demand was low, with an initial print run of 400 failing to sell out. Copernicus had made the book extremely technical, unreadable to all but the most advanced astronomers of the day, allowing it to disseminate into their ranks before stirring great controversy. And, like Osiander, contemporary mathematicians and astronomers encouraged its audience to view it as a useful mathematical fiction with no physical reality, thereby somewhat shielding it from accusations of blasphemy.

Among some astronomers, the book "at once took its place as a worthy successor to the *Almagest* of Ptolemy, which had hitherto been the Alpha and Omega of astronomers". Erasmus Reinhold hailed the work in 1542 and by 1551 had developed the *Prutenic Tables* ("Prussian Tables"; Latin: *Tabulae prutenicae*; German: *Preußische Tafeln*) using Copernicus' methods. The *Prutenic Tables*, published in 1551, were used as a basis for the calendar reform instituted in 1582 by Pope Gregory XIII. They were also used by sailors and maritime explorers, whose 15th-century predecessors had used Regiomontanus' *Table of the Stars*. In England, Robert Recorde, John Dee, Thomas Digges and William Gilbert were among those who adopted his position; in Germany, Christian Wurstisen, Christoph Rothmann and Michael Mästlin, the teacher of Johannes Kepler; in Italy, Giambattista Benedetti and Giordano Bruno whilst Franciscus Patricius accepted the rotation of the earth. In Spain, rules published in 1561 for the curriculum of the University of Salamanca gave students the choice between studying Ptolemy or Copernicus. One of those students, Diego de Zúñiga, published an acceptance of Copernican theory in 1584.

Very soon, nevertheless, Copernicus' theory was attacked with Scripture and with the common Aristotelian proofs. In 1549 Melanchthon, Luther's principal lieutenant, wrote against Copernicus, pointing to the theory's apparent conflict with Scripture and advocating that "severe measures" be taken to restrain the impiety of Copernicans. The works of Copernicus and Zúñiga—the latter for asserting that *De revolutionibus* was compatible with Catholic faith—were placed on the Index of Forbidden Books by a decree of the Sacred Congregation of March 5, 1616 (more than 70 years after Copernicus' publication):

This Holy Congregation has also learned about the spreading and acceptance by many of the false Pythagorean doctrine, altogether contrary to the Holy Scripture, that the earth moves and the sun is motionless, which is also taught by Nicholaus Copernicus' *De revolutionibus orbium coelestium* and by Diego de Zúñiga's *In Job* ... Therefore, in order that this opinion may not creep any further to the prejudice of Catholic truth, the Congregation has decided that the books by Nicolaus Copernicus [*De revolutionibus*] and Diego de Zúñiga [*In Job*] be suspended until corrected.

De revolutionibus was not formally banned but merely withdrawn from circulation, pending "corrections" that would clarify the theory's status as hypothesis. Nine sentences that represented the heliocentric system as certain were to be omitted or changed. After these corrections were prepared and formally approved in 1620 the reading of the book was permitted. But the book was never reprinted with the changes and was available in Catholic jurisdictions only to suitably qualified scholars, by special request. It remained on the Index until 1758, when Pope Benedict XIV (1740–58) removed the uncorrected book from his revised Index.

Census of Copies

Arthur Koestler described *De revolutionibus* as "*The Book That Nobody Read*" saying the book "was and is an all-time worst seller", despite the fact that it was reprinted four times. Owen Gingerich, a writer on both Nicolaus Copernicus and Johannes Kepler, disproved this after a 35-year project to examine every surviving copy of the first two editions. Gingerich showed that nearly all the leading mathematicians and astronomers of the time owned and read the book; however, his analysis of the marginalia shows that they almost all ignored the cosmology at the beginning of the book and were only interested in Copernicus' new equant-free models of planetary motion in the later chapters. Also, Nicolaus Reimers in 1587 translated the book into German.

Gingerich's efforts and conclusions are recounted in *The Book Nobody Read*, published in 2004 by Walker & Co. His census included 276 copies of the first edition (by comparison, there are 228 extant copies of the First Folio of Shakespeare) and 325 copies of the second. The research behind this book earned its author the Polish government's Order of Merit in 1981. Due largely to Gingerich's scholarship, *De revolutionibus* has been researched and catalogued better than any other first-edition historic text except for the original Gutenberg Bible. One of the copies now resides at the Archives of the University of Santo Tomas in the Miguel de Benavides Library. In January 2017, a second-edition copy was stolen as part of a heist of rare books from Heathrow Airport and remains unrecovered.

Editions

- 1543, Nuremberg, by Johannes Petreius.
- 1566, Basel, by Henricus Petrus.
- 1617, Amsterdam, by Nicolaus Mulerius.
- 1854, Warsaw, with Polish translation and the authentic preface by Copernicus.
- 1873, Thorn, German translation sponsored by the local *Coppernicus* Society, with all Copernicus' textual corrections given as footnotes.

Philosophiæ Naturalis Principia Mathematica

Philosophiæ Naturalis Principia Mathematica, often referred to as simply the *Principia*, is a work in three books by Isaac Newton, in Latin, first published 5 July 1687. After annotating and correcting his personal copy of the first edition, Newton published two further editions, in 1713 and 1726. The *Principia* states Newton's laws of motion, forming the foundation of classical mechanics; Newton's law of universal gravitation; and a derivation of Kepler's laws of planetary motion (which Kepler first obtained empirically). The *Principia* is considered as one of the most important works in the history of science.

The French mathematical physicist Alexis Clairaut assessed it in 1747: "The famous book of *Mathematical Principles of Natural Philosophy* marked the epoch of a great revolution in physics. The method followed by its illustrious author Sir Newton ... spread the light of mathematics on a science which up to then had remained in the darkness of conjectures and hypotheses." A more recent assessment has been that while acceptance of Newton's theories was not immediate, by the end of a century after publication in 1687, "no one could deny that" (out of the *Principia*) "a science had emerged that, at least in certain respects, so far exceeded anything that had ever gone before that it stood alone as the ultimate exemplar of science generally."

In formulating his physical theories, Newton developed and used mathematical methods now included in the field of calculus. But the language of calculus as we know it was largely absent from the *Principia*; Newton gave many of his proofs in a geometric form of infinitesimal calculus, based on limits of ratios of vanishing small geometric quantities. In a revised conclusion to the *Principia*, Newton used his expression that became famous, *Hypotheses non fingo* ("I formulate no hypotheses").

Expressed Aim and Topics Covered

Sir Isaac Newton (1643–1727) author of the *Principia*.

In the preface of the *Principia*, Newton wrote:

[...] Rational Mechanics will be the science of motions resulting from any forces whatsoever, and of the forces required to produce any motions, accurately proposed and demonstrated [...] And therefore we offer this work as mathematical principles of philosophy. For all the difficulty of philosophy seems to consist in this—from the phenomena of motions to investigate the forces of Nature, and then from these forces to demonstrate the other phenomena [...]

The *Principia* deals primarily with massive bodies in motion, initially under a variety of conditions and hypothetical laws of force in both non-resisting and resisting media, thus offering criteria to decide, by observations, which laws of force are operating in phenomena that may be observed. It attempts to cover hypothetical or possible motions both of celestial bodies and of terrestrial projectiles. It explores difficult problems of motions perturbed by multiple attractive forces. Its third and final book deals with the interpretation of observations about the movements of planets and their satellites. It shows how astronomical observations prove the inverse square law of gravitation (to an accuracy that was high by the standards of Newton's time); offers estimates of relative masses for the known giant planets and for the Earth and the Sun; defines the very slow motion of the Sun relative to the solar-system barycenter; shows how the theory of gravity can account for irregularities in the motion of the Moon; identifies the oblateness of the figure of the Earth; accounts approximately for marine tides including phenomena of spring and neap tides by the perturbing (and varying) gravitational attractions of the Sun and Moon on the Earth's waters; explains the precession of the equinoxes as an effect of the gravitational attraction of the Moon on the Earth's equatorial bulge; and gives theoretical basis for numerous phenomena about comets and their elongated, near-parabolic orbits.

The opening sections of the *Principia* contain, in revised and extended form, nearly all of the content of Newton's 1684 tract *De motu corporum in gyrum*.

The *Principia* begin with "Definitions" and "Axioms or Laws of Motion", and continues in three books.

Book 1, *De Motu Corporum*

Book 1, subtitled *De motu corporum* (*On the motion of bodies*) concerns motion in the absence of any resisting medium. It opens with a mathematical exposition of "the method of first and last ratios", a geometrical form of infinitesimal calculus.

The second section establishes relationships between centripetal forces and the law of areas now known as Kepler's second law (Propositions 1–3), and relates circular velocity and radius of path-curvature to radial force (Proposition 4), and relationships between centripetal forces varying as the inverse-square of the distance to the center and orbits of conic-section form (Propositions 5–10).

Propositions 11–31 establish properties of motion in paths of eccentric conic-section form including ellipses, and their relation with inverse-square central forces directed to a focus, and include Newton's theorem about ovals (lemma 28).

Propositions 43–45 are demonstration that in an eccentric orbit under centripetal force where the apse may move, a steady non-moving orientation of the line of apses is an indicator of an inverse-square law of force.

Book 1 contains some proofs with little connection to real-world dynamics. But there are also sections with far-reaching application to the solar system and universe:

Propositions 57–69 deal with the "motion of bodies drawn to one another by centripetal forces". This section is of primary interest for its application to the solar system, and includes Proposition 66 along with its 22 corollaries: here Newton took the first steps in the definition and study of the problem of the movements of three massive bodies subject to their mutually perturbing gravitational attractions, a problem which later gained name and fame (among other reasons, for its great difficulty) as the three-body problem.

Propositions 70–84 deal with the attractive forces of spherical bodies. The section contains Newton's proof that a massive spherically symmetrical body attracts other bodies outside itself as if all its mass were concentrated at its centre. This fundamental result, called the Shell theorem, enables the inverse square law of gravitation to be applied to the real solar system to a very close degree of approximation.

Book 2

Part of the contents originally planned for the first book was divided out into a second book, which largely concerns motion through resisting mediums. Just as Newton examined consequences of different conceivable laws of attraction in Book 1, here he examines different conceivable laws of resistance; thus Section 1 discusses resistance in direct proportion to velocity, and Section 2 goes on to examine the implications of resistance in proportion to the square of velocity. Book 2 also discusses (in Section 5) hydrostatics and the properties of compressible fluids. The effects of air resistance on pendulums are studied in Section 6, along with Newton's account of experiments that he carried out, to try to find out some characteristics of air resistance in reality by observing the motions of pendulums under different conditions. Newton compares the resistance offered by a medium against motions of globes with different properties (material, weight, size). In Section 8, he derives rules to determine the speed of waves in fluids and relates them to the density and condensation(Proposition 48; this would become very important in acoustics). He assumes that these rules apply equally to light and sound and estimates that the speed of sound is around 1088 feet per second and can increase depending on the amount of water in air.

Less of Book 2 has stood the test of time than of Books 1 and 3, and it has been said that Book 2 was largely written on purpose to refute a theory of Descartes which had some wide acceptance before Newton's work (and for some time after). According to this Cartesian theory of vortices, planetary motions were produced by the whirling of fluid vortices that filled interplanetary space and carried the planets along with them. Newton wrote at the end of Book 2 his conclusion that the hypothesis of vortices was completely at odds with the astronomical phenomena, and served not so much to explain as to confuse them.

Book 3, *De Mundi Systemate*

Book 3, subtitled *De mundi systemate* (*On the system of the world*), is an exposition of many consequences of universal gravitation, especially its consequences for astronomy. It builds upon the propositions of the previous books, and applies them with further specificity than in Book 1 to the motions observed in the solar system. Here (introduced by Proposition 22, and continuing in Propositions 25–35) are developed several of the features and irregularities of the orbital motion of the Moon, especially the variation. Newton lists the astronomical observations on which he relies, and establishes in a stepwise manner that the inverse square law of mutual gravitation applies to solar system bodies, starting with the satellites of Jupiter and going on by stages to show that the law is of universal application. He also gives starting at Lemma 4 and Proposition 40 the theory of the motions of comets, for which much data came from John Flamsteed and Edmond Halley, and accounts for the tides, attempting quantitative estimates of the contributions of the Sun and Moon to the tidal motions; and offers the first theory of the precession of the equinoxes. Book 3 also considers the harmonic oscillator in three dimensions, and motion in arbitrary force laws.

In Book 3 Newton also made clear his heliocentric view of the solar system, modified in a somewhat modern way, since already in the mid-1680s he recognised the "deviation of the Sun" from the centre of gravity of the solar system. For Newton, "the common centre of gravity of the Earth, the Sun and all the Planets is to be esteem'd the Centre of the World", and that this centre "either is at rest, or moves uniformly forward in a right line". Newton rejected the second alternative after adopting the position that "the centre of the system of the world is immoveable", which "is acknowledg'd by all, while some contend that the Earth, others, that the Sun is fix'd in that centre". Newton estimated the mass ratios Sun:Jupiter and Sun:Saturn, and pointed out that these put the centre of the Sun usually a little way off the common center of gravity, but only a little, the distance at most "would scarcely amount to one diameter of the Sun".

Commentary on the *Principia*

The sequence of definitions used in setting up dynamics in the *Principia* is recognisable in many textbooks today. Newton first set out the definition of mass[6].

The quantity of matter is that which arises conjointly from its density and magnitude. A body twice as dense in double the space is quadruple in quantity. This quantity I designate by the name of body or of mass.

This was then used to define the "quantity of motion" (today called momentum), and the principle of inertia in which mass replaces the previous Cartesian notion of *intrinsic force*. This then set the stage for the introduction of forces through the change in momentum of a body. Curiously, for today's readers, the exposition looks dimensionally incorrect, since Newton does not introduce the dimension of time in rates of changes of quantities.

He defined space and time "not as they are well known to all". Instead, he defined "true" time and space as "absolute" and explained:

Only I must observe, that the vulgar conceive those quantities under no other notions but from the relation they bear to perceptible objects. And it will be convenient to distinguish them into absolute and relative, true and apparent, mathematical and common. [...] instead of absolute places and motions, we use relative ones; and that without any inconvenience in common affairs; but in philosophical discussions, we ought to step back from our senses, and consider things themselves, distinct from what are only perceptible measures of them.

To some modern readers it can appear that some dynamical quantities recognised today were used in the *Principia* but not named. The mathematical aspects of the first two books were so clearly consistent that they were easily accepted; for example, Locke asked Huygens whether he could trust the mathematical proofs, and was assured about their correctness.

However, the concept of an attractive force acting at a distance received a cooler response. In his notes, Newton wrote that the inverse square law arose naturally due to the structure of matter. However, he retracted this sentence in the published version, where he stated that the motion of planets is consistent with an inverse square law, but refused to speculate on the origin of the law. Huygens and Leibniz noted that the law was incompatible with the notion of the aether. From a Cartesian point of view, therefore, this was a faulty theory. Newton's defence has been adopted since by many famous physicists—he pointed out that the mathematical form of the theory had to be correct since it explained the data, and he refused to speculate further on the basic nature of gravity. The sheer number of phenomena that could be organised by the theory was so impressive that younger "philosophers" soon adopted the methods and language of the *Principia.*

Rules of Reasoning in Philosophy

Perhaps to reduce the risk of public misunderstanding, Newton included at the beginning of Book 3 (in the second (1713) and third (1726) editions) a section entitled "Rules of Reasoning in Philosophy." In the four rules, as they came finally to stand in the 1726 edition, Newton effectively offers a methodology for handling unknown phenomena in nature and reaching towards explanations for them. The four Rules of the 1726 edition run as follows (omitting some explanatory comments that follow each):

Rule 1: *We are to admit no more causes of natural things than such as are both true and sufficient to explain their appearances.*

Rule 2: *Therefore to the same natural effects we must, as far as possible, assign the same causes.*

Rule 3: *The qualities of bodies, which admit neither intensification nor remission of degrees, and which are found to belong to all bodies within the reach of our experiments, are to be esteemed the universal qualities of all bodies whatsoever.*

Rule 4: *In experimental philosophy we are to look upon propositions inferred by general induction from phenomena as accurately or very nearly true, not withstanding any contrary hypothesis that may be imagined, till such time as other phenomena occur, by which they may either be made more accurate, or liable to exceptions.*

This section of Rules for philosophy is followed by a listing of 'Phenomena', in which are listed a number of mainly astronomical observations, that Newton used as the basis for inferences later on, as if adopting a consensus set of facts from the astronomers of his time.

Both the 'Rules' and the 'Phenomena' evolved from one edition of the *Principia* to the next. Rule 4 made its appearance in the third (1726) edition; Rules 1–3 were present as 'Rules' in the second (1713) edition, and predecessors of them were also present in the first edition of 1687, but there they had a different heading: they were not given as 'Rules', but rather in the first (1687) edition the predecessors of the three later 'Rules', and of most of the later 'Phenomena', were all lumped together under a single heading 'Hypotheses' (in which the third item was the predecessor of a heavy revision that gave the later Rule 3).

From this textual evolution, it appears that Newton wanted by the later headings 'Rules' and 'Phenomena' to clarify for his readers his view of the roles to be played by these various statements.

In the third (1726) edition of the *Principia*, Newton explains each rule in an alternative way and/or gives an example to back up what the rule is claiming. The first rule is explained as a philosophers' principle of economy. The second rule states that if one cause is assigned to a natural effect, then the same cause so far as possible must be assigned to natural effects of the same kind: for example respiration in humans and in animals, fires in the home and in the Sun, or the reflection of light whether it occurs terrestrially or from the planets. An extensive explanation is given of the third rule, concerning the qualities of bodies, and Newton discusses here the generalisation of observational results, with a caution against making up fancies contrary to experiments, and use of the rules to illustrate the observation of gravity and space.

Isaac Newton's statement of the four rules revolutionised the investigation of phenomena. With these rules, Newton could in principle begin to address all of the world's present unsolved mysteries. He was able to use his new analytical method to replace that of Aristotle, and he was able to use his method to tweak and update Galileo's experimental method. The re-creation of Galileo's method has never been significantly changed and in its substance, scientists use it today.

General Scholium

The *General Scholium* is a concluding essay added to the second edition, 1713 (and amended in the third edition, 1726). It is not to be confused with the *General Scholium* at the end of Book 2, Section 6, which discusses his pendulum experiments and resistance due to air, water, and other fluids.

Here Newton used what became his famous expression Hypotheses non fingo, "I formulate no hypotheses", in response to criticisms of the first edition of the *Principia*. (*'Fingo'* is sometimes nowadays translated 'feign' rather than the traditional 'frame'.) Newton's gravitational attraction, an invisible force able to act over vast distances, had led to criticism that he had introduced "occult agencies" into science. Newton firmly rejected such criticisms and wrote that it was enough that the phenomena implied gravitational attraction, as they did; but the phenomena did not so far indicate the cause of this gravity, and it was both unnecessary and improper to frame hypotheses of things not implied by the phenomena: such hypotheses "have no place in experimental philosophy", in contrast to the proper way in which "particular propositions are inferr'd from the phenomena and afterwards rendered general by induction".

Newton also underlined his criticism of the vortex theory of planetary motions, of Descartes, pointing to its incompatibility with the highly eccentric orbits of comets, which carry them "through all parts of the heavens indifferently".

Newton also gave theological argument. From the system of the world, he inferred the existence of a Lord God, along lines similar to what is sometimes called the argument from intelligent or purposive design. It has been suggested that Newton gave "an oblique argument for a unitarian conception of God and an implicit attack on the doctrine of the Trinity", but the General Scholium appears to say nothing specifically about these matters.

Halley and Newton's Initial Stimulus

In January 1684, Edmond Halley, Christopher Wren and Robert Hooke had a conversation in which Hooke claimed to not only have derived the inverse-square law, but also all the laws of planetary motion. Wren was unconvinced, Hooke did not produce the claimed derivation although the others gave him time to do it, and Halley, who could derive the inverse-square law for the restricted circular case (by substituting Kepler's relation into Huygens' formula for the centrifugal force) but failed to derive the relation generally, resolved to ask Newton.

Halley's visits to Newton in 1684 thus resulted from Halley's debates about planetary motion with Wren and Hooke, and they seem to have provided Newton with the incentive and spur to develop and write what became *Philosophiae Naturalis Principia Mathematica* (*Mathematical Principles of Natural Philosophy*). Halley was at that time a Fellow and Council member of the Royal Society in London (positions that in 1686 he resigned to become the Society's paid Clerk). Halley's visit to Newton in Cambridge in 1684 probably occurred in August. When Halley asked Newton's opinion on the problem of planetary motions discussed earlier that year between Halley, Hooke and Wren, Newton surprised Halley by saying that he had already made the derivations some time ago; but that he could not find the papers. (Matching accounts of this meeting come from Halley and Abraham De Moivre to whom Newton confided.) Halley then had to wait for Newton to 'find' the results, but in November 1684 Newton sent Halley an amplified version of whatever previous work Newton had done on the subject. This took the form of a 9-page manuscript, *De motu corporum in gyrum* (*Of the motion of bodies in an orbit*): the title is shown on some surviving copies, although the (lost) original may have been without title.

Newton's tract *De motu corporum in gyrum*, which he sent to Halley in late 1684, derived what are now known as the three laws of Kepler, assuming an inverse square law of force, and generalised the result to conic sections. It also extended the methodology by adding the solution of a problem on the motion of a body through a resisting medium. The contents of *De motu* so excited Halley by their mathematical and physical originality and far-reaching implications for astronomical theory, that he immediately went to visit Newton again, in November 1684, to ask Newton to let the Royal Society have more of such work. The results of their meetings clearly helped to stimulate Newton with the enthusiasm needed to take his investigations of mathematical problems much further in this area of physical science, and he did so in a period of highly concentrated work that lasted at least until mid-1686.

Newton's single-minded attention to his work generally, and to his project during this time, is shown by later reminiscences from his secretary and copyist of the period, Humphrey Newton. His account tells of Isaac Newton's absorption in his studies, how he sometimes forgot his food, or his sleep, or the state of his clothes, and how when he took a walk in his garden he would sometimes rush back to his room with some new thought, not even waiting to sit before beginning to write it down. Other evidence also shows Newton's absorption in the *Principia*: Newton for years kept up a regular programme of chemical or alchemical experiments, and he normally kept dated notes of them, but for a period from May 1684 to April 1686, Newton's chemical notebooks have no entries at all. So it seems that Newton abandoned pursuits to which he was normally dedicated, and did very little else for well over a year and a half, but concentrated on developing and writing what became his great work.

The first of the three constituent books was sent to Halley for the printer in spring 1686, and the other two books somewhat later. The complete work, published by Halley at his own financial risk, appeared in July 1687. Newton had also communicated *De motu* to Flamsteed, and during the period of composition he exchanged a few letters with Flamsteed about observational data on the planets, eventually acknowledging Flamsteed's contributions in the published version of the *Principia* of 1687.

Preliminary Version

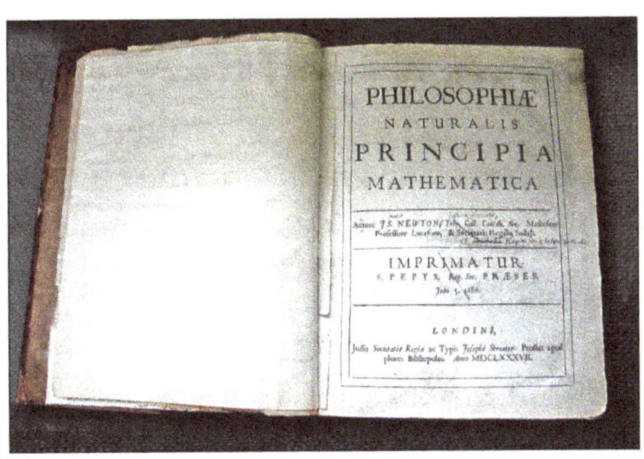

Newton's own first edition copy of his *Principia*, with handwritten corrections for the second edition.

The process of writing that first edition of the *Principia* went through several stages and drafts: some parts of the preliminary materials still survive, while others are lost except for fragments and cross-references in other documents.

Surviving materials show that Newton (up to some time in 1685) conceived his book as a two-volume work. The first volume was to be titled *De motu corporum, Liber primus*, with contents that later appeared in extended form as Book 1 of the *Principia*.

A fair-copy draft of Newton's planned second volume *De motu corporum, Liber secundus* survives, its completion dated to about the summer of 1685. It covers the application of the results of *Liber primus* to the Earth, the Moon, the tides, the solar system, and the universe; in this respect it has much the same purpose as the final Book 3 of the *Principia*, but it is written much less formally and is more easily read.

Titlepage and frontispiece of the third edition, London, 1726 (John Rylands Library).

It is not known just why Newton changed his mind so radically about the final form of what had been a readable narrative in *De motu corporum, Liber secundus* of 1685, but he largely started afresh in a new, tighter, and less accessible mathematical style, eventually to produce Book 3 of the *Principia* as we know it. Newton frankly admitted that this change of style was deliberate when he wrote that he had (first) composed this book "in a popular method, that it might be read by many", but to "prevent the disputes" by readers who could not "lay aside the[ir] prejudices", he had "reduced" it "into the form of propositions (in the mathematical way) which should be read by those only, who had first made themselves masters of the principles established in the preceding books". The final Book 3 also contained in addition some further important quantitative results arrived at by Newton in the meantime, especially about the theory of the motions of comets, and some of the perturbations of the motions of the Moon.

The result was numbered Book 3 of the *Principia* rather than Book 2, because in the meantime, drafts of *Liber primus* had expanded and Newton had divided it into two books. The new and final Book 2 was concerned largely with the motions of bodies through resisting mediums.

But the *Liber secundus* of 1685 can still be read today. Even after it was superseded by Book 3 of the *Principia*, it survived complete, in more than one manuscript. After Newton's death in 1727, the relatively accessible character of its writing encouraged the publication of an English translation in 1728 (by persons still unknown, not authorised by Newton's heirs). It appeared under the English title *A Treatise of the System of the World*. This had some amendments relative to Newton's manuscript of 1685, mostly to remove cross-references that used obsolete numbering to cite the propositions of an early draft of Book 1 of the *Principia*. Newton's heirs shortly afterwards published the Latin version in their possession, also in 1728, under the (new) title *De Mundi Systemate*, amended to update cross-references, citations and diagrams to those of the later editions of the *Principia*, making it look superficially as if it had been written by Newton after the *Principia*, rather than before. The *System of the World* was sufficiently popular to stimulate two revisions (with similar changes as in the Latin printing), a second edition (1731), and a 'corrected' reprint of the second edition (1740).

Halley's Role as Publisher

The text of the first of the three books of the *Principia* was presented to the Royal Society at the close of April 1686. Hooke made some priority claims (but failed to substantiate them), causing some delay. When Hooke's claim was made known to Newton, who hated disputes, Newton threatened to withdraw and suppress Book 3 altogether, but Halley, showing considerable diplomatic skills, tactfully persuaded Newton to withdraw his threat and let it go forward to publication. Samuel Pepys, as President, gave his imprimatur on 30 June 1686, licensing the book for publication. The Society had just spent its book budget on a *History of Fishes*, and the cost of publication was borne by Edmund Halley (who was also then acting as publisher of the *Philosophical Transactions of the Royal Society*): the book appeared in summer 1687.

Historical Context

Beginnings of the Scientific Revolution

Nicolaus Copernicus (1473–1543) was the first person to formulate a comprehensive heliocentric (or Sun-centered) model of the universe.

Nicolaus Copernicus had moved the Earth away from the center of the universe with the heliocentric theory for which he presented evidence in his book *De revolutionibus orbium coelestium* (*On the revolutions of the heavenly spheres*) published in 1543. The structure was completed when Johannes Kepler wrote the book *Astronomia nova* (*A new astronomy*) in 1609, setting out the evidence that planets move in elliptical orbits with the sun at one focus, and that planets do not move with constant speed along this orbit. Rather, their speed varies so that the line joining the centres of the sun and a planet sweeps out equal areas in equal times. To these two laws he added a third a decade later, in his book *Harmonices Mundi* (*Harmonies of the world*). This law sets out a proportionality between the third power of the characteristic distance of a planet from the sun and the square of the length of its year.

The foundation of modern dynamics was set out in Galileo's book *Dialogo sopra i due massimi sistemi del mondo* (*Dialogue on the two main world systems*) where the notion of inertia was implicit and used. In addition, Galileo's experiments with inclined planes had yielded precise mathematical relations between elapsed time and acceleration, velocity or distance for uniform and uniformly accelerated motion of bodies.

Descartes' book of 1644 *Principia philosophiae* (*Principles of philosophy*) stated that bodies can act on each other only through contact: a principle that induced people, among them himself, to hypothesize a universal medium as the carrier of interactions such as light and gravity—the aether. Newton was criticized for apparently introducing forces that acted at distance without any medium. Not until the development of particle theory was Descartes' notion vindicated when it was possible to describe all interactions, like the strong, weak, and electromagnetic fundamental interactions, using mediating gauge bosons and gravity through hypothesized gravitons. Although he was mistaken in his treatment of circular motion, this effort was more fruitful in the short term when it led others to identify circular motion as a problem raised by the principle of inertia. Christiaan Huygens solved this problem in the 1650s and published it much later in 1673 in his book *Horologium oscillatorium sive de motu pendulorum*.

Newton's Role

Newton had studied these books, or, in some cases, secondary sources based on them, and taken notes entitled *Quaestiones quaedam philosophicae* (*Questions about philosophy*) during his days as an undergraduate. During this period (1664–1666) he created the basis of calculus, and performed the first experiments in the optics of colour. At this time, his proof that white light was a combination of primary colours (found via prismatics) replaced the prevailing theory of colours and received an overwhelmingly favourable response, and occasioned bitter disputes with Robert Hooke and others, which forced him to sharpen his ideas to the point where he already composed sections of his later book *Opticks* by the 1670s in response. Work on calculus is shown in various papers and letters, including two to Leibniz. He became a fellow of the Royal Society and the second Lucasian Professor of Mathematics (succeeding Isaac Barrow) at Trinity College, Cambridge.

Newton's Early Work on Motion

In the 1660s Newton studied the motion of colliding bodies, and deduced that the centre of mass of two colliding bodies remains in uniform motion. Surviving manuscripts of the 1660s also show Newton's interest in planetary motion and that by 1669 he had shown, for a circular case of planetary motion, that the force he called 'endeavour to recede' (now called centrifugal force) had an inverse-square relation with distance from the center. After his 1679–1680 correspondence with Hooke, described below, Newton adopted the language of inward or centripetal force. According to Newton scholar J Bruce Brackenridge, although much has been made of the change in language and difference of point of view, as between centrifugal or centripetal forces, the actual computations and proofs remained the same either way. They also involved the combination of tangential and radial displacements, which Newton was making in the 1660s. The difference between the centrifugal and centripetal points of view, though a significant change of perspective, did not change the analysis. Newton also clearly expressed the concept of linear inertia in the 1660s: for this Newton was indebted to Descartes' work published 1644.

Controversy with Hooke

Hooke published his ideas about gravitation in the 1660s and again in 1674. He argued for an attracting principle of gravitation in *Micrographia* of 1665, in a 1666 Royal Society lecture *On gravity*, and again in 1674, when he published his ideas about the *System of the World* in somewhat developed form, as an addition to *An Attempt to Prove the Motion of the Earth from Observations.* Hooke clearly postulated mutual attractions between the Sun and planets, in a way that increased with nearness to the attracting body, along with a principle of linear inertia. Hooke's statements up to 1674 made no mention, however, that an inverse square law applies or might apply to these attractions. Hooke's gravitation was also not yet universal, though it approached universality more closely than previous hypotheses. Hooke also did not provide accompanying evidence or mathematical demonstration. On these two aspects, Hooke stated in 1674: "Now what these several degrees [of gravitational attraction] are I have not yet experimentally verified" (indicating that he did not yet know what law the gravitation might follow); and as to his whole proposal: "This I only hint at present", "having my self many other things in hand which I would first compleat, and therefore cannot so well attend it" (i.e., "prosecuting this Inquiry").

Imaginary portrait of English polymath Robert Hooke (1635–1703).

In November 1679, Hooke began an exchange of letters with Newton, of which the full text is now published. Hooke told Newton that Hooke had been appointed to manage the Royal Society's correspondence, and wished to hear from members about their researches, or their views about the researches of others; and as if to whet Newton's interest, he asked what Newton thought about various matters, giving a whole list, mentioning "compounding the celestial motions of the planets of a direct motion by the tangent and an attractive motion towards the central body", and "my hypothesis of the lawes or causes of springinesse", and then a new hypothesis from Paris about planetary motions (which Hooke described at length), and then efforts to carry out or improve national surveys, the difference of latitude between London and Cambridge, and other items. Newton's reply offered "a fansy of my own" about a terrestrial experiment (not a proposal about celestial motions) which might detect the Earth's motion, by the use of a body first suspended in air and then dropped to let it fall. The main point was to indicate how Newton thought the falling body could experimentally reveal the Earth's motion by its direction of deviation from the vertical, but he went on hypothetically to consider how its motion could continue if the solid Earth had not been in the way (on a spiral path to the centre). Hooke disagreed with Newton's idea of how the body would continue to move. A short further correspondence developed, and towards the end of it Hooke, writing on 6 January 1680 to Newton, communicated his "supposition ... that the Attraction always is in a duplicate proportion to the Distance from the Center Reciprocall, and Consequently that the Velocity will be in a subduplicate proportion to the Attraction and Consequently as Kepler Supposes Reciprocall to the Distance." (Hooke's inference about the velocity was actually incorrect.)

In 1686, when the first book of Newton's *Principia* was presented to the Royal Society, Hooke claimed that Newton had obtained from him the "notion" of "the rule of the decrease of Gravity, being reciprocally as the squares of the distances from the Center". At the same time (according to Edmond Halley's contemporary report) Hooke agreed that "the Demonstration of the Curves generated therby" was wholly Newton's.

A recent assessment about the early history of the inverse square law is that "by the late 1660s," the assumption of an "inverse proportion between gravity and the square of distance was rather common and had been advanced by a number of different people for different reasons". Newton himself had shown in the 1660s that for planetary motion under a circular assumption, force in the radial direction had an inverse-square relation with distance from the center. Newton, faced in May 1686 with Hooke's claim on the inverse square law, denied that Hooke was to be credited as author of the idea, giving reasons including the citation of prior work by others before Hooke. Newton also firmly claimed that even if it had happened that he had first heard of the inverse square proportion from Hooke, which it had not, he would still have some rights to it in view of his mathematical developments and demonstrations, which enabled observations to be relied on as evidence of its accuracy, while Hooke, without mathematical demonstrations and evidence in favour of the supposition, could only guess (according to Newton) that it was approximately valid "at great distances from the center".

The background described above shows there was basis for Newton to deny deriving the inverse square law from Hooke. On the other hand, Newton did accept and acknowledge, in all editions of the *Principia*, that Hooke (but not exclusively Hooke) had separately appreciated the inverse square law in the solar system. Newton acknowledged Wren, Hooke and Halley in this connection in the Scholium to Proposition 4 in Book 1. Newton also acknowledged to Halley that his correspondence with Hooke in 1679–80 had reawakened his dormant interest in astronomical matters, but that did not mean, according to Newton, that Hooke had told Newton anything new or original: "yet am I not beholden to him for any light into that business but only for the diversion he gave me from my other studies to think on these things & for his dogmaticalness in writing as if he had found the motion in the Ellipsis, which inclined me to try it ...".) Newton's reawakening interest in astronomy received further stimulus by the appearance of a comet in the winter of 1680/1681, on which he corresponded with John Flamsteed.

In 1759, decades after the deaths of both Newton and Hooke, Alexis Clairaut, mathematical astronomer eminent in his own right in the field of gravitational studies, made his assessment after reviewing what Hooke had published on gravitation. "One must not think that this idea ... of Hooke diminishes Newton's glory", Clairaut wrote; "The example of Hooke" serves "to show what a distance there is between a truth that is glimpsed and a truth that is demonstrated".

Location of Early-edition Copies

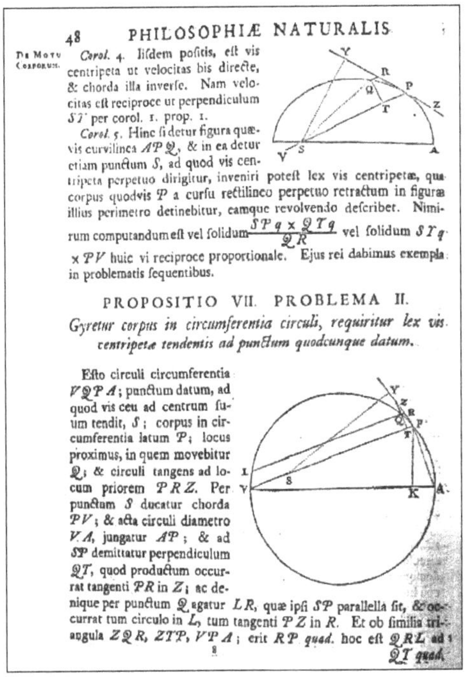

48 PHILOSOPHIÆ NATURALIS

De Motu Corporum.

Corol. 4. Iisdem positis, est vis centripeta ut velocitas bis directe, & chorda illa inverse. Nam velocitas est reciproce ut perpendiculum *ST* per corol. 1. prop. 1.

Corol. 5. Hinc si detur figura quævis curvilinea *APQ*, & in ea detur etiam punctum *S*, ad quod vis centripeta perpetuo dirigitur, inveniri potest lex vis centripetæ, qua corpus quodvis *P* a cursu rectilineo perpetuo retractum in figuræ illius perimetro detinebitur, eamque revolvendo describet. Nimirum computandum est vel solidum $\frac{SPq \times QTq}{QR}$ vel solidum $STq \times PV$ huic vi reciproce proportionale. Ejus rei dabimus exempla in problematis sequentibus.

PROPOSITIO VII. PROBLEMA II.

Gyretur corpus in circumferentia circuli, requiritur lex vis centripetæ tendentis ad punctum quodcunque datum.

Esto circuli circumferentia *VQPA*; punctum datum, ad quod vis ceu ad centrum suum tendit, *S*; corpus in circumferentia latum *P*; locus proximus, in quem movebitur *Q*; & circuli tangens ad locum priorem *PRZ*. Per punctum *S* ducatur chorda *PV*; & acta circuli diametro *VA*, jungatur *AP*; & ad *SP* demittatur perpendiculum *QT*, quod productum occurrat tangenti *PR* in *Z*; ac denique per punctum *Q* agatur *LR*, quæ ipsi *SP* parallela sit, & occurrat tum circulo in *L*, tum tangenti *PZ* in *R*. Et ob similia triangula *ZQR*, *ZTP*, *VPA*; erit *RP quad.* hoc est *QRL* ad *QT quad.*

A page from the *Principia*.

Since only between 250 and 400 copies were printed by the Royal Society, the first edition is very rare. Several rare-book collections contain first edition and other early copies of Newton's *Principia Mathematica*, including:

- Cambridge University Library has Newton's own copy of the first edition, with handwritten notes for the second edition.
- The Earl Gregg Swem Library at the College of William & Mary has a first edition copy of the Principia. In it are notes in Latin throughout by a not yet identified hand.
- The Frederick E. Brasch Collection of Newton and Newtoniana in Stanford University also has a first edition of the Principia.
- A first edition forms part of the Crawford Collection, housed at the Royal Observatory, Edinburgh.
- The Uppsala University Library owns a first edition copy, which was stolen in the 1960s and returned to the library in 2009.
- The Folger Shakespeare Library in Washington, D.C. owns a first edition, as well as a 1713 second edition.
- The Huntington Library in San Marino, California owns Isaac Newton's personal copy, with annotations in Newton's own hand.
- The Martin Bodmer Library keeps a copy of the original edition that was owned by Leibniz. In it, we can see handwritten notes by Leibniz, in particular concerning the controversy of

who first formulated calculus (although he published it later, Newton argued that he developed it earlier).

In 2016, a first edition sold for $3.7 million.

A facsimile edition (based on the 3rd edition of 1726 but with variant readings from earlier editions and important annotations) was published in 1972 by Alexandre Koyré and I. Bernard Cohen.

Later Editions

Two later editions were published by Newton:

Second Edition, 1713

Newton had been urged to make a new edition of the *Principia* since the early 1690s, partly because copies of the first edition had already become very rare and expensive within a few years after 1687. Newton referred to his plans for a second edition in correspondence with Flamsteed in November 1694: Newton also maintained annotated copies of the first edition specially bound up with interleaves on which he could note his revisions; two of these copies still survive: but he had not completed the revisions by 1708, and of two would-be editors, Newton had almost severed connections with one, Nicolas Fatio de Duillier, and the other, David Gregory seems not to have met with Newton's approval and was also terminally ill, dying later in 1708. Nevertheless, reasons were accumulating not to put off the new edition any longer. Richard Bentley, master of Trinity College, persuaded Newton to allow him to undertake a second edition, and in June 1708 Bentley wrote to Newton with a specimen print of the first sheet, at the same time expressing the (unfulfilled) hope that Newton had made progress towards finishing the revisions. It seems that Bentley then realised that the editorship was technically too difficult for him, and with Newton's consent he appointed Roger Cotes, Plumian professor of astronomy at Trinity, to undertake the editorship for him as a kind of deputy (but Bentley still made the publishing arrangements and had the financial responsibility and profit). The correspondence of 1709–1713 shows Cotes reporting to two masters, Bentley and Newton, and managing (and often correcting) a large and important set of revisions to which Newton sometimes could not give his full attention. Under the weight of Cotes' efforts, but impeded by priority disputes between Newton and Leibniz, and by troubles at the Mint, Cotes was able to announce publication to Newton on 30 June 1713. Bentley sent Newton only six presentation copies; Cotes was unpaid; Newton omitted any acknowledgement to Cotes.

Among those who gave Newton corrections for the Second Edition were: Firmin Abauzit, Roger Cotes and David Gregory. However, Newton omitted acknowledgements to some because of the priority disputes. John Flamsteed, the Astronomer Royal, suffered this especially.

Third Edition, 1726

The third edition was published 25 March 1726, under the stewardship of *Henry Pemberton, M.D., a man of the greatest skill in these matters...*; Pemberton later said that this recognition was worth more to him than the two hundred guinea award from Newton.

Annotated and Other Editions

In 1739–42, two French priests, Pères Thomas LeSeur and François Jacquier (of the Minim order, but sometimes erroneously identified as Jesuits), produced with the assistance of J.-L. Calandrini an extensively annotated version of the *Principia* in the 3rd edition of 1726. Sometimes this is referred to as the *Jesuit edition*: it was much used, and reprinted more than once in Scotland during the 19th century.

Émilie du Châtelet also made a translation of Newton's Principia into French. Unlike LeSeur and Jacquier's edition, hers was a complete translation of Newton's three books and their prefaces. She also included a Commentary section where she fused the three books into a much clearer and easier to understand summary. She included an analytical section where she applied the new mathematics of calculus to Newton's most controversial theories. Previously, geometry was the standard mathematics used to analyse theories. Du Châtelet's translation is the only complete one to have been done in French and hers remains the standard French translation to this day.

English Translations

Two full English translations of Newton's *Principia* have appeared, both based on Newton's 3rd edition of 1726.

The first, from 1729, by Andrew Motte, was described by Newton scholar I. Bernard Cohen (in 1968) as "still of enormous value in conveying to us the sense of Newton's words in their own time, and it is generally faithful to the original: clear, and well written". The 1729 version was the basis for several republications, often incorporating revisions, among them a widely used modernised English version of 1934, which appeared under the editorial name of Florian Cajori (though completed and published only some years after his death). Cohen pointed out ways in which the 18th-century terminology and punctuation of the 1729 translation might be confusing to modern readers, but he also made severe criticisms of the 1934 modernised English version, and showed that the revisions had been made without regard to the original, also demonstrating gross errors "that provided the final impetus to our decision to produce a wholly new translation".

The second full English translation, into modern English, is the work that resulted from this decision by collaborating translators I. Bernard Cohen, Anne Whitman, and Julia Budenz; it was published in 1999 with a guide by way of introduction.

William H. Donahue has published a translation of the work's central argument, published in 1996, along with expansion of included proofs and ample commentary. The book was developed as a textbook for classes at St. John's College and the aim of this translation is to be faithful to the Latin text.

Homages

In 2014, British astronaut Tim Peake named his upcoming mission to the International Space Station *Principia* after the book, in "honour of Britain's greatest scientist". Tim Peake's *Principia* launched on December 15, 2015 aboard Soyuz TMA-19M.

References

- Cong Ma; Tongjie Zhang (2010). "Power of Observational Hubble Parameter Data: a Figure of Merit Exploration". Astrophysical Journal. 730 (2): 74. Bibcode:2011ApJ...730...74M. arXiv:1007.3787. doi:10.1088/0004-637X/730/2/74
- Overbye, Dennis (17 March 2014). "Detection of Waves in Space Buttresses Landmark Theory of Big Bang". New York Times. Retrieved 05, January 2020
- MP Hobson; GP Efstathiou; AN Lasenby (2006). General Relativity: An introduction for physicists (Reprinted with corrections 2007 ed.). Cambridge University Press. p. 187. ISBN 978-0-521-82951-9
- Ghose, Tia (26 February 2015). "Big Bang, Deflated? Universe May Have Had No Beginning". Live Science. Retrieved 20, March 2020
- Vilenkin, Alex (2007). Many worlds in one : the search for other universes. New York: Hill and Wang, A division of Farrar, Straus and Giroux. p. 19. ISBN 978-0-8090-6722-0
- See M. Sami; R. Myrzakulov (2015). "Late time cosmic acceleration: ABCD of dark energy and modified theories of gravity". International Journal of Modern Physics D. 25 (12). arXiv:1309.4188. doi:10.1142/S0218271816300317
- Beringer, J.; et al. (Particle Data Group) (2012). "2013 Review of Particle Physics" (PDF). Phys. Rev. D. 86: 010001. Bibcode:2012PhRvD..86a0001B. doi:10.1103/PhysRevD.86.010001
- Ali, Ahmed Faraq (4 February 2015). "Cosmology from quantum potential". Physics Letters B. 741: 276–279. doi:10.1016/j.physletb.2014.12.057. Retrieved 12, February 2020
- Exirifard, Q. (2010). "Phenomenological covariant approach to gravity". General Relativity and Gravitation. 43: 93–106. Bibcode:2011GReGr..43...93E. arXiv:0808.1962. doi:10.1007/s10714-010-1073-6
- Overbye, Dennis (June 15, 2016). "Scientists Hear a Second Chirp From Colliding Black Holes". The New York Times. Retrieved 07, June 2020
- Peebles, P. J. E.; Ratra, Bharat (2003). "The cosmological constant and dark energy". Reviews of Modern Physics. 75 (2): 559–606. Bibcode:2003RvMP...75..559P. arXiv:astro-ph/0207347. doi:10.1103/RevModPhys.75.559
- Overbye, Dennis (20 February 2017). "Cosmos Controversy: The Universe Is Expanding, but How Fast?". New York Times. Retrieved 19, August 2020
- "The Newest Search for Gravitational Waves has Begun". LIGO Caltech. LIGO. 18 September 2015. Retrieved 26, July 2020
- Zhong-Yue Wang (2016). "Modern Theory for Electromagnetic Metamaterials". Plasmonics. 11 (2): 503–508. doi:10.1007/s11468-015-0071-7
- Weiss, Achim. "Equilibrium and change: The physics behind Big Bang Nucleosynthesis". Einstein Online. Archived from the original on 8 February 2007. Retrieved 19, May 2020

5

Types of Dark Matter

Dark matter is composed of particles that are not detectable by observing electromagnetic radiations because they do not reflect, absorb or emit light. This chapter discusses in detail cold dark matter, hot dark matter, mixed dark matter and dark matter halo. These topics are crucial for a complete understanding of the subject.

In some clusters, the space between galaxies is filled with gas so hot, scientists cannot see it using visible light telescopes. The gas only can be seen as X-rays or gamma rays. Scientists look at that gas and measure how much there is between galaxies in clusters. By doing this, they discovered that there must be five times more material in the clusters than we can detect. The invisible matter that we can't detect is called "dark matter."

The Fermi Gamma-Ray Space Telescope can detect high-energy gamma rays that may be emitted when dark matter particles collide.

Perhaps the most significant sign of the existence of dark matter, however, is our very existence. Despite its invisibility, dark matter has been critical to the evolution of our universe and to the emergence of stars, planets and even life. That is because dark matter carries five times the mass of ordinary matter and, furthermore, does not directly interact with light. Both these properties were critical to the creation of structures such as galaxies—within the (relatively short) time span we know to be a typical galaxy lifetime—and, in particular, to the formation of a galaxy the size of the Milky Way. Without dark matter, radiation would have prevented clumping of the galactic structure for too long, in essence wiping it out and keeping the universe smooth and homogeneous. The galaxy essential to our solar system and our life was formed in the time since the big bang only because of the existence of dark matter.

Dark matter possibly could be brown dwarfs, "failed" stars that never ignited because they lacked the mass needed to start burning. Dark matter could be white dwarfs, the remnants of cores of dead small- to medium-size stars. Or dark matter could be neutron stars or black holes, the remnants of large stars after they explode.

However, problems exist with each of these suggestions. Scientists have strong evidence there aren't enough brown dwarfs or white dwarfs to account for all the dark matter. Black holes and neutron stars, too, are rare.

Dark matter may not be made up of the matter we are familiar with at all. The matter that makes up dark matter could different. It may be filled with particles predicted by theory but that scientists have yet to observe.

Because scientists can't see dark matter directly, they have found other ways to investigate it. We can use indirect ways to study things, like looking at a shadow and making an educated guess about what's casting the shadow. One way scientists indirectly study dark matter is by using gravitational lensing.

Light going through a gravitational lens is similar to light going through an optical lens: It gets bent. When light from distant stars passes through a galaxy or cluster, the gravity of the matter present in the galaxy or cluster causes the light to bend. As a result, the light looks like it is coming from somewhere else rather than from its actual origin. The amount of bending helps scientists learn about the dark matter present. Many NASA scientists use the Hubble Space Telescope to observe gravitational lensing.

In addition to these indirect ways, scientists at NASA think they have a direct way to detect dark matter using the Fermi Gamma-Ray Space Telescope. This telescope looks at gamma rays, the highest energy form of light. When two dark matter particles crash into each other, they might release a gamma ray. The Fermi Telescope could theoretically detect these collisions, which would appear as a burst of a gamma ray in the sky.

Technical Definition

In standard cosmology, matter is anything whose energy density scales with the inverse cube of the scale factor, i.e., $\rho \propto a^{-3}$. This is in contrast to radiation, which scales as the inverse fourth power of the scale factor $\rho \propto a^{-4}$, and a cosmological constant, which is independent of a. These scalings can be understood intuitively: for an ordinary particle in a cubical box, doubling the length of the sides of the box decreases the density (and hence energy density) by a factor of eight (2^3). For radiation, the decrease in energy density is larger because an increase in scale factor causes a proportional redshift. A cosmological constant, as an intrinsic property of space, has a constant energy density regardless of the volume under consideration.

In principle, "dark matter" means all components of the universe that are not visible but still obey $\rho \propto a^{-3}$. In practice, the term "dark matter" is often used to mean only the non-baryonic component of dark matter, i.e., excluding "missing baryons." Context will usually indicate which meaning is intended.

Observational Evidence

Galaxy Rotation Curves

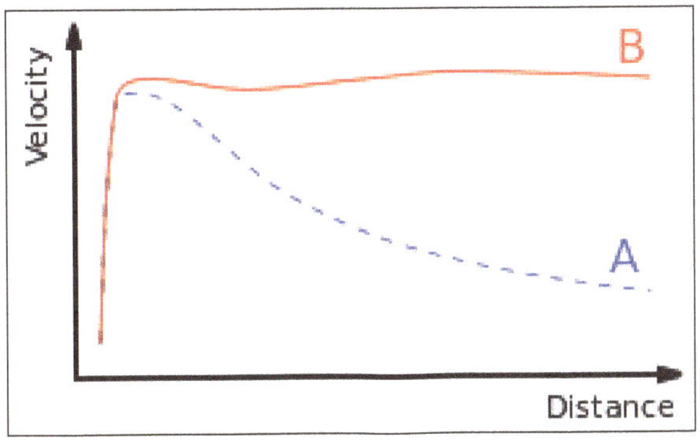

Rotation curve of a typical spiral galaxy: predicted (**A**) and observed (**B**). Dark matter can explain the 'flat' appearance of the velocity curve out to a large radius.

The arms of spiral galaxies rotate around the galactic center. The luminous mass density of a spiral galaxy decreases as one goes from the center to the outskirts. If luminous mass were all the matter, then we can model the galaxy as a point mass in the centre and test masses orbiting around it, similar to the solar system. From Kepler's Second Law, it is expected that the rotation velocities will decrease with distance from the center, similar to the Solar System. This is not observed.Instead, the galaxy rotation curve remains flat as distance from the center increases.

If Kepler's laws are correct, then the obvious way to resolve this discrepancy is to conclude that the mass distribution in spiral galaxies is not similar to that of the Solar System. In particular, there is a lot of non-luminous matter (dark matter) in the outskirts of the galaxy.

Velocity Dispersions

Stars in bound systems must obey the virial theorem. The theorem, together with the measured velocity distribution, can be used to measure the mass distribution in a bound system, such as elliptical galaxies or globular clusters. With some exceptions, velocity dispersion estimates of elliptical galaxies do not match the predicted velocity dispersion from the observed mass distribution, even assuming complicated distributions of stellar orbits.

As with galaxy rotation curves, the obvious way to resolve the discrepancy is to postulate the existence of non-luminous matter.

Galaxy Clusters

Galaxy clusters are particularly important for dark matter studies since their masses can be estimated in three independent ways:

- From the scatter in radial velocities of the galaxies within clusters.
- From X-rays emitted by hot gas in the clusters. From the X-ray energy spectrum and flux, the gas temperature and density can be estimated, hence giving the pressure; assuming pressure and gravity balance determines the cluster›s mass profile.

- Gravitational lensing (usually of more distant galaxies) can measure cluster masses without relying on observations of dynamics (e.g., velocity).

Generally, these three methods are in reasonable agreement that dark matter outweighs visible matter by approximately 5 to 1.

Strong gravitational lensing as observed by the Hubble Space Telescope in Abell indicates the presence of dark matter.

Gravitational Lensing

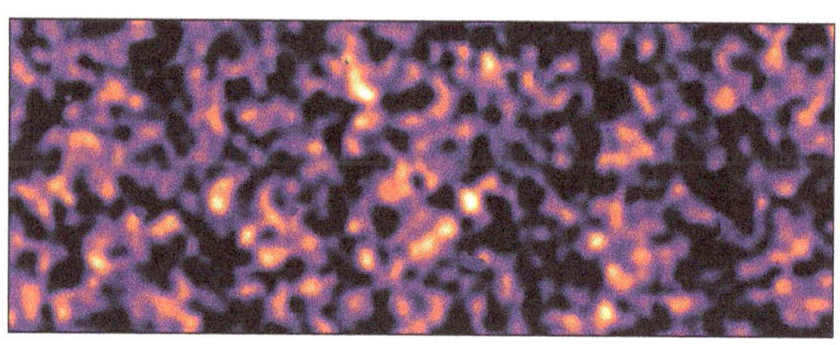

Dark matter map for a patch of sky based on gravitational lensing analysis of a Kilo-Degree survey.

One of the consequences of general relativity is that massive objects (such as a cluster of galaxies) lying between a more distant source (such as a quasar) and an observer should act as a lens to bend the light from this source. The more massive an object, the more lensing is observed.

Strong lensing is the observed distortion of background galaxies into arcs when their light passes through such a gravitational lens. It has been observed around many distant clusters including Abell 1689. By measuring the distortion geometry, the mass of the intervening cluster can be obtained. In the dozens of cases where this has been done, the mass-to-light ratios obtained correspond to the dynamical dark matter measurements of clusters. Lensing can lead to multiple copies of an image. By analyzing the distribution of multiple image copies, scientists have been able to deduce and map the distribution of dark matter around the MACS J0416.1-2403 galaxy cluster.

Weak gravitational lensing investigates minute distortions of galaxies, using statistical analyses from vast galaxy surveys. By examining the apparent shear deformation of the adjacent background galaxies, the mean distribution of dark matter can be characterized. The mass-to-light ratios correspond to dark matter densities predicted by other large-scale structure measurements. Dark matter does not bend light itself; mass (in this case the mass of the dark matter) bends spacetime. Light follows the curvature of spacetime, resulting in the lensing effect.

Cosmic Microwave Background

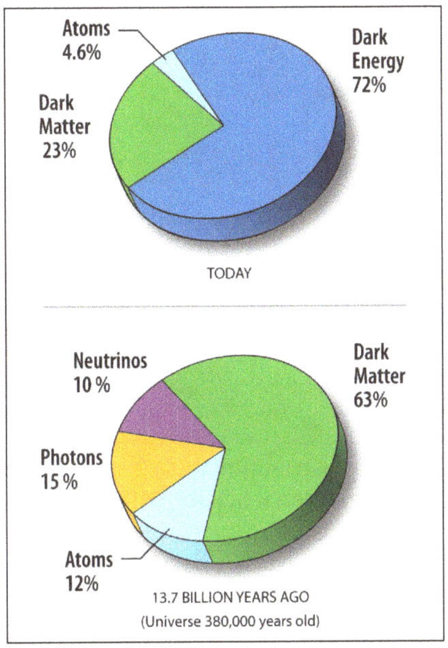

Estimated division of total energy in the universe into matter, dark matter and dark energy based on five years of WMAP data.

Although both dark matter and ordinary matter are matter, they do not behave in the same way. In particular, in the early universe, ordinary matter was ionized and interacted strongly with radiation via Thomson scattering. Dark matter does not interact directly with radiation, but it does affect the CMB by its gravitational potential (mainly on large scales), and by its effects on the density and velocity of ordinary matter. Ordinary and dark matter perturbations, therefore, evolve differently with time and leave different imprints on the cosmic microwave background (CMB).

The cosmic microwave background is very close to a perfect blackbody but contains very small temperature anisotropies of a few parts in 100,000. A sky map of anisotropies can be decomposed into an angular power spectrum, which is observed to contain a series of acoustic peaks at near-equal spacing but different heights. The series of peaks can be predicted for any assumed set of cosmological parameters by modern computer codes such as CMBFast and CAMB, and matching theory to data, therefore, constrains cosmological parameters. The first peak mostly shows the density of baryonic matter, while the third peak relates mostly to the density of dark matter, measuring the density of matter and the density of atoms.

The CMB anisotropy was first discovered by COBE in 1992, though this had too coarse resolution to detect the acoustic peaks. After the discovery of the first acoustic peak by the balloon-borne BOOMERanGexperiment in 2000, the power spectrum was precisely observed by WMAP in 2003-12, and even more precisely by the Planck spacecraft in 2013-15. The results support the Lambda-CDM model.

The observed CMB angular power spectrum provides powerful evidence in support of dark matter, as its precise structure is well fitted by the Lambda-CDM model but difficult to reproduce with any competing model such as MOND.

Structure Formation

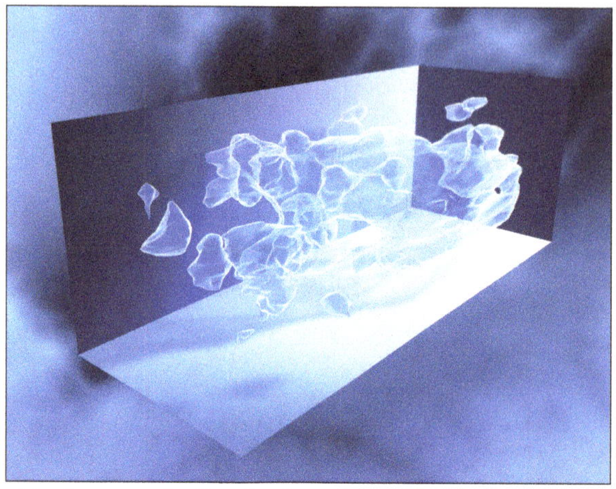

3D map of the large-scale distribution of dark matter, reconstructed from measurements of weak gravitational lensing with the Hubble Space Telescope.

Structure formation refers to the period after the Big Bang when density perturbations collapsed to form stars, galaxies, and clusters. Prior to structure formation, the Friedmann solutions to general relativity describe a homogeneous universe. Later, small anisotropies gradually grew and condensed the homogeneous universe into stars, galaxies and larger structures. Ordinary matter is affected by radiation, which is the dominant element of the universe at very early times. As a result, its density perturbations are washed out and unable to condense into structure. If there were only ordinary matter in the universe, there would not have been enough time for density perturbations to grow into the galaxies and clusters currently seen.

Dark matter provides a solution to this problem because it is unaffected by radiation. Therefore, its density perturbations can grow first. The resulting gravitational potential acts as an attractive potential well for ordinary matter collapsing later, speeding up the structure formation process.

Bullet Cluster

If dark matter does not exist, then the next most likely explanation is that general relativity – the prevailing theory of gravity – is incorrect. The Bullet Cluster, the result of a recent collision of two galaxy clusters, provides a challenge for modified gravity theories because its apparent center of mass is far displaced from the baryonic center of mass. Standard dark matter theory can easily explain this observation, but modified gravity has a much harder time, especially since the observational evidence is model-independent.

Type Ia Supernova Distance Measurements

Type Ia supernovae can be used as standard candles to measure extragalactic distances, which can in turn be used to measure how fast the universe has expanded in the past. The data indicates that the universe is expanding at an accelerating rate, the cause of which is usually ascribed to dark energy. Since observations indicate the universe is almost flat, it is expected that the total energy density of everything in the universe to sum to 1 ($\Omega_{tot} \sim 1$). The measured dark energy density is $\Omega_{\Lambda} = \sim 0.690$; the observed ordinary (baryonic) matter energy density is $\Omega_b = \sim 0.0482$ and the energy density of radiation is negligible. This leaves a missing $\Omega_{dm} = \sim 0.258$ that nonetheless behaves like matter – dark matter.

Sky Surveys and Baryon Acoustic Oscillations

Baryon acoustic oscillations (BAO) are regular, periodic fluctuations in the density of the visible baryonic matter (normal matter) of the universe. These are predicted to arise in the Lambda-CDM model due to the early universe›s acoustic oscillations in the photon-baryon fluid and can be observed in the cosmic microwave background angular power spectrum. BAOs set up a preferred length scale for baryons. As the dark matter and baryons clumped together after recombination, the effect is much weaker in the galaxy distribution in the nearby universe, but is detectable as a subtle (~ 1 percent) preference for pairs of galaxies to be separated by 147 Mpc, compared to those separated by 130 or 160 Mpc. This feature was predicted theoretically in the 1990s and then discovered in 2005, in two large galaxy redshift surveys, the Sloan Digital Sky Survey and the 2dF Galaxy Redshift Survey. Combining the CMB observations with BAO measurements from galaxy redshift surveys provides a precise estimate of the Hubble constant and the average matter density in the Universe. The results support the Lambda-CDM model.

Redshift-space Distortions

Large galaxy redshift surveys may be used to make a three-dimensional map of the galaxy distribution. These maps are slightly distorted because distances are estimated from observed redshifts; the redshift contains a contribution from the galaxy's so-called peculiar velocity in addition to the dominant Hubble expansion term. On average, superclusters are expanding but more slowly than the cosmic mean due to their gravity, while voids are expanding faster than average. In a redshift map, galaxies in front of a supercluster have excess radial velocities towards it and have redshifts slightly higher than their distance would imply, while galaxies behind the supercluster have redshifts slightly low for their distance. This effect causes superclusters to appear squashed in the radial direction, and likewise voids are stretched. Their angular positions are unaffected. The effect is not detectable for any one structure since the true shape is not known, but can be measured by averaging over many structures assuming Earth is not at a special location in the Universe.

The effect was predicted quantitatively by Nick Kaiser in 1987, and first decisively measured in 2001 by the 2dF Galaxy Redshift Survey. Results are in agreement with the Lambda-CDM model.

Lyman-alpha Forest

In astronomical spectroscopy, the Lyman-alpha forest is the sum of the absorption lines arising from the Lyman-alphatransition of neutral hydrogen in the spectra of distant galaxies and quasars. Lyman-alpha forest observations can also constrain cosmological models. These constraints agree with those obtained from WMAP data.

Composition of Dark Matter: Baryonic vs. Nonbaryonic

Dark matter can refer to any substance that interacts predominantly via gravity with visible matter (e.g., stars and planets). Hence in principle it need not be composed of a new type of fundamental particle but could, at least in part, be made up of standard baryonic matter, such as protons or neutrons. However, for the reasons outlined below, most scientists think the dark matter is dominated by a non-baryonic component, which is likely composed of a currently unknown fundamental particle (or similar exotic state).

Baryonic Matter

Baryons (protons and neutrons) make up ordinary stars and planets. However, baryonic matter also encompasses less common black holes, neutron stars, faint old white dwarfs and brown dwarfs, collectively known as massive compact halo objects(MACHOs), which can be hard to detect.

However multiple lines of evidence suggest the majority of dark matter is not made of baryons:

- Sufficient diffuse, baryonic gas or dust would be visible when backlit by stars.
- The theory of Big Bang nucleosynthesis predicts the observed abundance of the chemical elements. If there are more baryons, then there should also be more helium, lithium and heavier elements synthesized during the Big Bang. Agreement with observed abundances requires that baryonic matter makes up between 4–5% of the universe›s critical density. In contrast, large-scale structure and other observations indicate that the total matter density is about 30% of the critical density.
- Astronomical searches for gravitational microlensing in the Milky Way found that at most a small fraction of the dark matter may be in dark, compact, conventional objects (MACHOs, etc.); the excluded range of object masses is from half the Earth›s mass up to 30 solar masses, which covers nearly all the plausible candidates.
- Detailed analysis of the small irregularities (anisotropies) in the cosmic microwave background. Observations by WMAP and Planck indicate that around five-sixths of the total matter is in a form that interacts significantly with ordinary matter or photons only through gravitational effects.

Non-baryonic Matter

Candidates for non-baryonic dark matter are hypothetical particles such as axions, sterile neutrinos, weakly-interacting massive particles (WIMPs), gravitationally-interacting massive particles (GIMPs), or supersymmetric particles. The three neutrino types already observed are indeed abundant, and dark, and matter, but because their individual masses – however uncertain they may be – are almost certainly tiny, they can only supply a small fraction of dark matter, due to limits derived from large-scale structure and high-redshift galaxies.

Unlike baryonic matter, nonbaryonic matter did not contribute to the formation of the elements in the early universe (Big Bang nucleosynthesis) and so its presence is revealed only via its gravitational effects, or, weak lensing. In addition, if the particles of which it is composed are supersymmetric, they can undergo annihilation interactions with themselves, possibly resulting in observable by-products such as gamma rays and neutrinos (indirect detection).

Dark Matter Aggregation and Dense Dark Matter Objects

If dark matter is as common as observations suggest, an obvious question is whether it can form objects equivalent to planets, stars, or black holes. The answer has historically been that it cannot, because of two factors:

- *It lacks an efficient means to lose energy:* Ordinary matter forms dense objects because it has numerous ways to lose energy. Losing energy would be essential for object formation, because a particle that gains energy during compaction or falling «inward» under gravity, and cannot lose it any other way, will heat up and increase velocity and momentum. Dark matter appears to lack means to lose energy, simply because it is not capable of interacting strongly in other ways except through gravity. The Virial theorem suggests that such a particle would not stay bound to the gradually forming object - as the object began to form and compact, the dark matter particles within it would speed up and tend to escape.
- *It lacks a range of interactions needed to form structures:* Ordinary matter interacts in many different ways. This allow it to form more complex structures. For example, stars form through gravity, but the particles within them interact and can emit energy in the form of neutrinos and electromagnetic radiation through fusion when they become energetic enough. Protons and neutrons can bind via the strong interaction and then form atoms with electrons largely through electromagnetic interaction. But there is no evidence that dark matter is capable of such a wide variety of interactions, since it only seems to interact through gravity and through some means no stronger than the weak interaction. (although this is speculative until dark matter is better understood).

This question has been debated heavily during recent years. In 2016–2017 the idea of dense dark matter or dark matter being black holes, including primordial black holes, made a comeback following results of gravitation wave detection. These were again ruled out in December 2017, but research and theories based on these still continue as at 2018, including approaches to dark matter cooling, and the question is by no means settled.

Classification of Dark Matter: Cold, Warm or Hot

Dark matter can be divided into *cold*, *warm*, and *hot* categories. These categories refer to velocity rather than an actual temperature, indicating how far corresponding objects moved due to random motions in the early universe, before they slowed due to cosmic expansion – this is an important distance called the free streaming length (FSL). Primordial density fluctuations smaller than this length get washed out as particles spread from overdense to underdense regions, while larger fluctuations are unaffected; therefore this length sets a minimum scale for later structure formation. The categories are set with respect to the size of a protogalaxy (an object that later evolves into a dwarf galaxy): dark matter particles are classified as cold, warm, or hot according as their FSL; much smaller (cold), similar (warm), or much larger (hot) than a protogalaxy.

Mixtures of the above are also possible: a theory of mixed dark matter was popular in the mid-1990s, but was rejected following the discovery of dark energy.

Cold dark matter leads to a bottom-up formation of structure while hot dark matter would result in a top-down formation scenario; the latter is excluded by high-redshift galaxy observations.

Alternative Definitions

These categories also correspond to fluctuation spectrum effects and the interval following the Big Bang at which each type became non-relativistic. Davis *et al.* wrote in 1985:

> Candidate particles can be grouped into three categories on the basis of their effect on the fluctuation spectrum(Bond *et al.* 1983). If the dark matter is composed of abundant light particles which remain relativistic until shortly before recombination, then it may be termed "hot". The best candidate for hot dark matter is a neutrino. A second possibility is for the dark matter particles to interact more weakly than neutrinos, to be less abundant, and to have a mass of order 1 keV. Such particles are termed "warm dark matter", because they have lower thermal velocities than massive neutrinos there are at present few candidate particles which fit this description. Gravitinos and photinos have been suggested (Pagels and Primack 1982; Bond, Szalay and Turner 1982). Any particles which became nonrelativistic very early, and so were able to diffuse a negligible distance, are termed "cold" dark matter (CDM). There are many candidates for CDM including supersymmetric particles.
>
> —*M. Davis, G. Efstathiou, C. S. Frenk, and S. D. M. White*

Another approximate dividing line is that warm dark matter became non-relativistic when the universe was approximately 1 year old and 1 millionth of its present size and in the radiation-dominated era (photons and neutrinos), with a photon temperature 2.7 million K. Standard physical cosmology gives the particle horizon size as 2*ct* (speed of light multiplied by time) in the radiation-dominated era, thus 2 light-years. A region of this size would expand to 2 million light years today (absent structure formation). The actual FSL is approximately 5 times the above length, since it continues to grow slowly as particle velocities decrease inversely with the scale factor after they become non-relativistic. In this example the FSL would correspond to 10 million light-years or 3 Mpc today, around the size containing an average large galaxy.

The 2.7 million K photon temperature gives a typical photon energy of 250 electron-volts, thereby setting a typical mass scale for warm dark matter: particles much more massive than this, such as GeV – TeV mass WIMPs, would become non-relativistic much earlier than 1 year after the Big Bang and thus have FSLs much smaller than a protogalaxy, making them cold. Conversely, much lighter particles, such as neutrinos with masses of only a few eV, have FSLs much larger than a protogalaxy, thus qualifying them as hot.

Cold Dark Matter

Cold dark matter offers the simplest explanation for most cosmological observations. It is dark matter composed of constituents with an FSL much smaller than a protogalaxy. This is the focus for dark matter research, as hot dark matter does not seem capable of supporting galaxy or galaxy cluster formation, and most particle candidates slowed early.

The constituents of cold dark matter are unknown. Possibilities range from large objects like MACHOs (such as black holes) or RAMBOs (such as clusters of brown dwarfs), to new particles such as WIMPs and axions.

Studies of Big Bang nucleosynthesis and gravitational lensing convinced most cosmologists that MACHOs cannot make up more than a small fraction of dark matter. According to A. Peter: "... the only *really plausible* dark-matter candidates are new particles."

The 1997 DAMA/NaI experiment and its successor DAMA/LIBRA in 2013, claimed to directly detect dark matter particles passing through the Earth, but many researchers remain skeptical, as negative results from similar experiments seem incompatible with the DAMA results.

Many supersymmetric models offer dark matter candidates in the form of the WIMPy Lightest Supersymmetric Particle(LSP). Separately, heavy sterile neutrinos exist in non-supersymmetric extensions to the standard model that explain the small neutrino mass through the seesaw mechanism.

Warm Dark Matter

Warm dark matter comprises particles with an FSL comparable to the size of a protogalaxy. Predictions based on warm dark matter are similar to those for cold dark matter on large scales, but with less small-scale density perturbations. This reduces the predicted abundance of dwarf galaxies and may lead to lower density of dark matter in the central parts of large galaxies. Some researchers consider this a better fit to observations. A challenge for this model is the lack of particle candidates with the required mass ~ 300 eV to 3000 eV.

No known particles can be categorized as warm dark matter. A postulated candidate is the sterile neutrino: a heavier, slower form of neutrino that does not interact through the weak force, unlike other neutrinos. Some modified gravity theories, such as scalar-tensor-vector gravity, require "warm" dark matter to make their equations work.

Hot Dark Matter

Hot dark matter consists of particles whose FSL is much larger than the size of a protogalaxy. The neutrino qualifies as such particle. They were discovered independently, long before the hunt for dark matter: they were postulated in 1930, and detected in 1956. Neutrinos' mass is less than 10^{-6} that of an electron. Neutrinos interact with normal matter only via gravity and the weak force, making them difficult to detect (the weak force only works over a small distance, thus a neutrino triggers a weak force event only if it hits a nucleus head-on). This makes them 'weakly-interacting light particles' (WILPs), as opposed to WIMPs.

The three known flavours of neutrinos are the *electron*, *muon*, and *tau*. Their masses are slightly different. Neutrinos oscillate among the flavours as they move. It is hard to determine an exact upper bound on the collective average mass of the three neutrinos (or for any of the three individually). For example, if the average neutrino mass were over 50 eV/c^2 (less than $10^{-5}$of the mass of an electron), the universe would collapse. CMB data and other methods indicate that their average mass probably does not exceed 0.3 eV/c^2. Thus, observed neutrinos cannot explain dark matter.

Because galaxy-size density fluctuations get washed out by free-streaming, hot dark matter implies that the first objects that can form are huge supercluster-size pancakes, which then fragment into galaxies. Deep-field observations show instead that galaxies formed first, followed by clusters and superclusters as galaxies clump together.

Detection of Dark Matter Particles

If dark matter is made up of sub-atomic particles, then millions, possibly billions, of such particles must pass through every square centimeter of the Earth each second. Many experiments aim to test this hypothesis. Although WIMPs are popular search candidates, the Axion Dark Matter eXperiment (ADMX) searches for axions. Another candidate is heavy hidden sector particles that only interact with ordinary matter via gravity.

These experiments can be divided into two classes: direct detection experiments, which search for the scattering of dark matter particles off atomic nuclei within a detector; and indirect detection, which look for the products of dark matter particle annihilations or decays.

Direct Detection

Direct detection experiments aim to observe low-energy recoils (typically a few keVs) of nuclei induced by interactions with particles of dark matter, which (in theory) are passing through the Earth. After such a recoil the nucleus will emit energy as, e.g., scintillation light or phonons, which is then detected by sensitive apparatus. To do this effectively it is crucial to maintain a low background, and so such experiments operate deep underground to reduce the interference from cosmic rays. Examples of underground laboratories with direct detection experiments include the Stawell mine, the Soudan mine, the SNOLAB underground laboratory at Sudbury, the Gran Sasso National Laboratory, the Canfranc Underground Laboratory, the Boulby Underground Laboratory, the Deep Underground Science and Engineering Laboratory and the China Jinping Underground Laboratory.

These experiments mostly use either cryogenic or noble liquid detector technologies. Cryogenic detectors operating at temperatures below 100mK, detect the heat produced when a particle hits an atom in a crystal absorber such as germanium. Noble liquid detectors detect scintillation produced by a particle collision in liquid xenon or argon. Cryogenic detector experiments include: CDMS, CRESST, EDELWEISS, EURECA. Noble liquid experiments include ZEPLIN, XENON, DEAP, ArDM, WARP, DarkSide, PandaX, and LUX, the Large Underground Xenon experiment. Both of these techniques focus strongly on their ability to distinguish background particles (which predominantly scatter off electrons) from dark matter particles (that scatter off nuclei). Other experiments include SIMPLE and PICASSO.

Currently there has been no well-established claim of dark matter detection from a direct detection experiment, leading instead to strong upper limits on the mass and interaction cross section with nucleons of such dark matter particles. The DAMA/NaI and more recent DAMA/LIBRA experimental collaborations claim to have detected an annual modulation in the rate of events in their detectors, which they claim is due to dark matter. This results from the expectation that as the Earth orbits the Sun, the velocity of the detector relative to the dark matter halo will vary by a small amount. This claim is so far unconfirmed and in contradiction with negative results from other experiments such as LUX and SuperCDMS.

A special case of direct detection experiments covers those with directional sensitivity. This is a search strategy based on the motion of the Solar System around the Galactic Center. A low pressure time projection chamber makes it possible to access information on recoiling tracks and constrain WIMP-nucleus kinematics. WIMPs coming from the direction in which the Sun travels (approximately towards Cygnus) may then be separated from background, which should be isotropic. Directional dark matter experiments include DMTPC, DRIFT, Newage and MIMAC.

Indirect Detection

Collage of six cluster collisions with dark matter maps. The clusters were observed in a study of how dark matter in clusters of galaxies behaves when the clusters collide.

Indirect detection experiments search for the products of the self-annihilation or decay of dark matter particles in outer space. For example, in regions of high dark matter density (e.g., the centre of our galaxy) two dark matter particles could annihilate to produce gamma rays or Standard Model particle-antiparticle pairs.Alternatively if the dark matter particle is unstable, it could decay into standard model (or other) particles. These processes could be detected indirectly through an excess of gamma rays, antiprotons or positrons emanating from high density regions in our galaxy or others. A major difficulty inherent in such searches is that various astrophysical sources can mimic the signal expected from dark matter, and so multiple signals are likely required for a conclusive discovery.

A few of the dark matter particles passing through the Sun or Earth may scatter off atoms and lose energy. Thus dark matter may accumulate at the center of these bodies, increasing the chance of collision/annihilation. This could produce a distinctive signal in the form of high-energy neutrinos. Such a signal would be strong indirect proof of WIMP dark matter. High-energy neutrino telescopes such as AMANDA, IceCube and ANTARES are searching for this signal. The detection by LIGO in September 2015 of gravitational waves, opens the possibility of observing dark matter in a new way, particularly if it is in the form of primordial black holes.

Many experimental searches have been undertaken to look for such emission from dark matter annihilation or decay, examples of which follow. The Energetic Gamma Ray Experiment Telescope observed more gamma rays in 2008 than expected from the Milky Way, but scientists concluded that this was most likely due to incorrect estimation of the telescope›s sensitivity.

The Fermi Gamma-ray Space Telescope is searching for similar gamma rays. In April 2012, an analysis of previously available data from its Large Area Telescope instrument produced statistical evidence of a 130 GeV signal in the gamma radiation coming from the center of the Milky Way. WIMP annihilation was seen as the most probable explanation.

At higher energies, ground-based gamma-ray telescopes have set limits on the annihilation of dark matter in dwarf spheroidal galaxies and in clusters of galaxies.

The PAMELA experiment (launched 2006) detected excess positrons. They could be from dark matter annihilation or from pulsars. No excess antiprotons were observed.

In 2013 results from the Alpha Magnetic Spectrometer on the International Space Station indicated excess high-energy cosmic rays that could be due to dark matter annihilation.

Collider Searches for Dark Matter

An alternative approach to the detection of dark matter particles in nature is to produce them in a laboratory. Experiments with the Large Hadron Collider (LHC) may be able to detect dark matter particles produced in collisions of the LHC protonbeams. Because a dark matter particle should have negligible interactions with normal visible matter, it may be detected indirectly as (large amounts of) missing energy and momentum that escape the detectors, provided other (non-negligible) collision products are detected. Constraints on dark matter also exist from the LEP experiment using a similar principle, but probing the interaction of dark matter particles with electrons rather than quarks. It is important to note that any discovery from collider searches must be corroborated by discoveries in the indirect or direct detection sectors to prove that the particle discovered is, in fact, the dark matter of our Universe.

Alternative Hypotheses

Because dark matter remains to be conclusively identified, many other hypotheses have emerged aiming to explain the observational phenomena that dark matter was conceived to explain. The most common method is to modify general relativity. General relativity is well-tested on solar system scales, but its validity on galactic or cosmological scales has not been well proven. A suitable modification to general relativity can conceivably eliminate the need for dark matter. The most well-known theories of this class are MOND and its relativistic generalization TeVeS, f(R) gravity and entropic gravity. Alternative theories abound.

A problem with alternative hypotheses is that the observational evidence for dark matter comes from so many independent approaches. Explaining any individual observation is possible but explaining all of them is very difficult. Nonetheless, there have been some scattered successes for alternative hypotheses, such as a 2016 test of gravitational lensing in entropic gravity.

The prevailing opinion among most astrophysicists is that while modifications to general relativity can conceivably explain part of the observational evidence, there is probably enough data to conclude there must be some form of dark matter.

Baryonic Dark Matter

The properties of a fermionic dark matter candidate which has baryon number and refer to this type of scenario as “Baryonic Dark Matter.” We show the relic density constraints and the predictions for direct detection experiments. Since this model has only four free parameters one could hope to test this idea combining the possible results from the Large Hadron Collider and dark matter experiments.

Recently, it proposed a simple extension of the Standard Model where one can understand the spontaneous breaking of baryon and lepton numbers at the low scale. Here, will be discussed a simplified version of this model, only considering baryon number as a local gauge symmetry. Therefore, this model is based on the gauge group,

$$SU(3)_C \otimes SU(2)_L \otimes U(1)_Y \otimes U(1)_B.$$

In order to define an anomaly-free theory using this gauge group, we need to include additional fermions that account for anomaly cancellation,

$$\Psi_L \sim (1, 2, -\frac{1}{2}, B_1), \qquad \Psi_R \sim (1, 2, -\frac{1}{2}, B_2),$$

$$\eta_R \sim (1, 1, -1, B_1), \qquad \eta_L \sim (1, 1, -1, B_2),$$

$$\chi_R \sim (1, 1, 0, B_1), \qquad \chi_L \sim (1, 1, 0, B_2),$$

and extend the scalar sector with a new Higgs boson to allow for a spontaneous breaking of baryon number,

$$S_B \sim (1, 1, 0, -3).$$

Here B_1 and B_2 refer to the baryon numbers of the additional fermions which are vector-like under the SM gauge group. From the conditions that ensure the cancellation of all relevant baryonic anomalies, one finds the following relation for the baryon numbers of the new fermions:

$$B_1 - B_2 = -3$$

The relevant interactions of the new fields in the theory are,

$$-\mathcal{L} \supset \lambda_1 \overline{\Psi}_L \Psi_R S_B + \lambda_2 \overline{\eta}_R \eta_L S_B + \lambda_3 \overline{\chi}_R \chi_L S_B + h.c.$$

Notice that one can have terms such as $a_1 \chi_L \chi_L S_B$ and $a_2 \chi_R \chi_R S_B$ † B only when $B_1 = -B_2$. Here we will stick to the case where $B_1 \neq -B_2$. The Yukawa interactions between the new fields and the Standard Model Higgs boson are present as well, but they are not relevant for our main discussion. It is important to notice that the new Higgs boson must have baryon number −3 in order to generate vector-like mass for the new fermions. Therefore, once S_B gets a vacuum expectation value breaking local $U(1)_B$ we never generate any operator mediating proton decay and the scale for baryon number violation can be as low as a few TeV.

In this simple theory, when the local baryon number is spontaneously broken by the vacuum expectation value v_B of S_B, one obtains a Z_2 discrete symmetry which protects the stability of the dark matter candidate. Under this Z_2 the new fermionic fields transform as,

$$\Psi_{L,R} \to -\Psi_{L,R}, \ \eta_{L,R} \to -\eta_{L,R}, \text{ and } \chi_{L,R} \to -\chi_{L,R}.$$

Therefore, when the lightest new field with baryon number is neutral, one can have a candidate for the cold dark matter in the Universe. For simplicity, we will focus on the case when the dark matter is SM singlet-like and is the Dirac fermion $\chi = \chi_R + \chi_L$. Since the dark matter has baryon number, the relevant interactions with the new gauge boson Z_B related to baryon number are,

$$\mathcal{L} \supset gB\,\overline{\chi}\gamma_\mu Z_B^\mu (B_2\, P_L + B_1\, P_R)\chi,$$

where P_L and P_R are the left- and right-handed projectors, and g_B is the gauge coupling related to baryon number. Of course, the new gauge boson also couples to the SM quarks, which is crucial to understand the properties of the dark matter candidate. The leptophobic gauge boson mass reads as,

$$M_{Z_B} = 3_{gB} v_B,$$

while the mass of the SM singlet-like baryonic dark matter candidate is given by,

$$M_\chi = \lambda_3 v_B / \sqrt{2} < \frac{\sqrt{2\pi}}{3} \frac{M_{Z_B}}{g_B}$$

This upper limit is coming from the perturbative condition on the Yukawa coupling λ_3, i.e. $|\lambda_3|^2/4\pi < 1$. It is important to notice that this model for baryonic dark matter has only four free parameters:

$$M_\chi, M_{Z_B}, g_B, \textit{and } B,$$

and one needs to satisfy the relic density constraints and the bounds from direct detection. Here $B = B_1 + B_2$ is the parameter which enters in the predictions for the relevant cross sections. One could imagine that the parameters M_{Z_B} and g_B can be determined from the discovery of a leptophobic gauge boson at the Large Hadron Collider. Therefore, one can say that for a given value of these two quantities we can find the values of B and M_χ using the relic density and spin-independent cross section values. We will discuss in more details the numerical predictions in the following sections in order to appreciate this connection between collider physics and dark matter experiments. In the rest of the paper we will neglect the kinetic mixing between $U(1)_B$ and $U(1)_Y$, as well as the mixing between the SM Higgs and S_B.

Non-Baryonic Dark Matter

Non-baryonic matter is matter that, unlike all the kind kinds of matter with which we are familiar, is not made of baryons (including the neutrons and protons found in all atomic nuclei). Proposed as a possible constituent of dark matter, it could come in two forms, classified as cold non-baryonic matter or hot non-baryonic matter. The former would be made of particles moving much slower than light, of which there are several, as yet undetected, candidates; hot non-baryonic matter would be made of particles moving very fast, such as neutrinos. Non-baryonic matter (hot or cold) is supposed to interact weakly with radiation. Therefore, the imprints left by the non-baryonic matter in the cosmic background radiation would be different than those left by the baryonic matter. This attribute could be used to measure the contribution of non-baryonic matter to the total amount of mass in the universe.

Cold Dark Matter

CDM has long been the leading candidate for what this missing mass in the universe is. However, some researchers still favor a combination theory, where aspects of all three types of dark matter exist together to make up the total missing mass.

CDM is a kind of dark matter that, if it exists, moves slowly compared to the speed of light. It is thought to have been present in the universe since the very beginning and has very likely influenced the growth and evolution of galaxies. as well as the formation of the first stars. Astronomers and physicists think that it's most likely some exotic particle that hasn't yet been detected. It very likely has some very specific properties:

It would have to lack an interaction with the electromagnetic force. This is fairly obvious, since dark matter is dark. Therefore it doesn't interact with, reflect, or radiate any type of energy in the electromagnetic spectrum.

However, any candidate particle that makes up cold dark matter would have to take into account that it has to interact with a gravitational field. For proof of this, astronomers have noticed that dark matter accumulations in galaxy clusters wield a gravitational influence on light from more distant objects that happens to be passing by. This so-called "gravitational lensing effect" has been observed many times.

Candidate Cold Dark Matter Objects

While no known matter meets all of the criteria for cold dark matter, at least three theories have been advanced to explain CDM.

- *Weakly Interacting Massive Particles*: Also known as WIMPs, these particles, by definition, meet all the needs of CDM. However, no such particle has ever been found to exist. WIMPs have become the catch all term for all cold dark matter candidates, regardless of why the particle is thought to arise.
- *Axions*: These particles possess (at least marginally) the necessary properties of dark matter, but for various reasons are probably not the answer to the question of cold dark matter.
- *MACHOs*: This is an acronym for *Massive Compact Halo Objects*, which are objects like black holes, ancient neutron stars, brown dwarfs and planetary objects. These are all non-luminous and massive. But, because of their large sizes, both in terms of volume and mass, they would be relatively easy to detect by monitoring localized gravitational interactions. However, there are problems with the MACHO hypothesis. The observed motion of galaxies, for instance, is uniform in a way that would be hard to explain if MACHOs supplied the missing mass. Furthermore, star clusters would require a very uniform distribution of

such objects within their boundaries. That seems very unlikely. Also, the sheer number of MACHOs that would have to be fairly large in order to explain the missing mass.

Structure Formation

In the cold dark matter theory, structure grows hierarchically, with small objects collapsing under their self-gravity first and merging in a continuous hierarchy to form larger and more massive objects. In the hot dark matter paradigm, popular in the early 1980s, structure does not form hierarchically (*bottom-up*), but rather forms by fragmentation (*top-down*), with the largest superclusters forming first in flat pancake-like sheets and subsequently fragmenting into smaller pieces like our galaxy the Milky Way. Predictions of the cold dark matter paradigm are in general agreement with observations of cosmological large scale structure.

Lambda CDM Model

Since the late 1980s or 1990s, most cosmologists favor the cold dark matter theory (specifically the modern Lambda-CDM model) as a description of how the Universe went from a smooth initial state at early times to the lumpy distribution of galaxies and their clusters we see today — the large-scale structure of the Universe. The theory sees the role that dwarf galaxies played as crucial, as they are thought to be natural building blocks that form larger structures, created by small-scale density fluctuations in the early Universe.

Hot Dark Matter

Hot DM refers to particles, such as neutrinos, that were moving at nearly the speed of light at redshift $z \sim 10^6$ (or time t ~ 1 yr), when the temperature $T \sim 3 \times 10^2$ eV and the cosmic horizon first encompassed 10^{12} $M\odot$,the amount of dark matter contained in the halo of a large galaxy like the Milky Way. Hot DM particles must also be still in thermal equilibrium after the last phase transition in the hot early universe, the QCD confinement transition, which presumably took place at $T_{QCD} \approx 10^2$ MeV. Hot DM particles have a cosmological number density roughly comparable to that of the microwave background photons, which implies an upper bound to their mass of a few tens of eV. This then implies that free streaming of these relativistic particles destroys any fluctuations smaller than supercluster size, $\sim 10^{15}$ $M\odot$.

The main problem with HDM theory is that the high speeds of the particles (i.e. neutrinos) in the early universe could not have allowed small density fluctuations to clump together in order to create the large fluctuations we see now. We believe matter (or in other words galaxies) is distributed throughout the universe as it is now due to the growth of small initial fluctuations. Since neutrinos would have been moving so fast that these tiny initial fluctuations would have been smoothed out, HDM theory cannot account for the distribution of galaxies in the universe. The small scale of this initial clumping that is impossible for neutrinos to maintain is supported by COBE observations.

Galaxy Formation with Hot DM

The standard hot DM candidate is massive neutrinos, although other, more exotic theoretical possibilities have been suggested, such as a "majoron" of nonzero mass which is lighter than the lightest neutrino species, and into which all neutrinos decay. Neutrinos appeared to be an attractive DM candidate because of the measurement of an electron neutrino mass of about 30 eV in 1980 . This coincided with the improving CMB limits on the primordial fluctuation amplitude, which forced Zel'dovich and other theorists to abandon the idea that all the dark matter could be made of ordinary baryonic matter. The version of HDM that they worked out in detail, with adiabatic Gaussian primordial fluctuations, became the prototype for the subsequent $\Omega_m = 1$CDM theory.

Mass Constraints

Direct measurements of neutrino masses have given only upper limits. A secure upper limit on the electron neutrino mass is roughly 15 eV. The Particle Data Group notes that a more precise limit cannot be given since unexplained effects have resulted in significantly negative measurements of $m(\nu_e)^2$ in tritium beta decay experiments. However, this problem is at least partially resolved, and the latest experimental upper limits on the electron neutrino mass are 2.8 eV from the Mainz and 2.5 eV from the Troitsk tritium beta decay experiments (both 95% C.L.). There is an upper limit on an effective Majorana neutrino mass of ~ 1eV from neutrinoless double beta decay experiments . The upper limits from accelerator experiments on the masses of the other neutrinos are $m(\nu_\mu) < 0.17\,\text{MeV}\,(90\%\,\text{CL})$ and $m(\nu_\tau) < 18\,\text{MeV}\,(95\%\,\text{CL})$, but since stable neutrinos with such large masses would certainly "overclose the universe" (i.e., contribute such a large cosmological density that the universe could never have attained its present age), cosmology implies a much lower upper limit on these neutrino masses.

Before going further, it will be necessary to discuss the thermal history of neutrinos in the standard hot big bang cosmology in order to derive the corresponding constraints on their mass. Left-handed neutrinos of mass $\leq 1\,\text{MeV}$ remain in thermal equilibrium until the temperature drops to $T_{\nu d}$, at which point their mean free path first exceeds the horizon size and they essentially cease interacting thereafter, except gravitationally. Their mean free path is, in natural units $(\hbar = c = 1)$, $\lambda_\nu \sim [\sigma_\nu n_{e\pm}]^{-1} [(G_F^2 T^2)(T^3)]^{-1}$, where $G_F \approx 10^{-5}\,G_e\,\text{V}^{-2}$ is the Fermi constant that measures the strength of the weak interactions. The horizon size is $\lambda_\nu \sim (G\rho)^{-1/2} \sim M_{P\ell} T^{-2}$,

the Planck mass $M_{P\ell} \equiv G^{-1/2} = 1.22 \times 10^{19}\text{GeV}$ Thus $\lambda_h / \lambda_\nu \sim (T/T_{\nu d})^3$, with the neutrino decoupling temperature,

$$T_{\nu d} \sim M_{P\ell}^{-1/3} G_F^{-1/3} \sim 1\,\text{MeV}.$$

After T drops below $\frac{1}{2}\,\text{MeV}$, e^+e^- annihilation ceases to be balanced by pair creation, and the entropy of the e^+e^- pairs heats the photons. Above $1\,\text{MeV}$, the number density $n_{\nu,i}$ of each left-handed neutrino species and its right-handed antiparticle is equal to that of the photons, n_γ, times the factor 3/4 from Fermi versus Bose statistics. But then e^+e^- annihilation increases the photon number density relative to that of the neutrinos by a factor of 11/4. As a result, the neutrino temperature $T_{\nu,0} = (4/11)^{1/3}\,T_{\gamma,0}$. Thus today, for each species,

$$n_{\nu,0} = \frac{3}{4} \cdot \frac{4}{11} n_{\gamma,0} = 109\,\theta^3\ \mathrm{cm}^{-3},$$

where $\theta \equiv (T_0/2.7\,\mathrm{K})$. With the cosmic background radiation temperature $T_0 = 2.728 \pm 0.004$ K measured by the FIRAS instrument on the COBE satellite, $T_{\nu,0} = 1.947$ K and $n_{\nu,0} = 112\,\mathrm{cm}^{-3}$.

Since the present cosmological matter density is,

$$\bar{\rho}m = \Omega\rho_c = 10.54\,\Omega_m h^2\ \mathrm{ke\,V\,cm}^{-3},$$

it follows that,

$$\sum_i m_{\nu i} < \bar{\rho}m / n_{\nu,0} \le 96\,\Omega_m h^2 \theta^{-3}\ \mathrm{eV} \approx 93\Omega_m h^2\ \mathrm{eV},$$

where the sum runs over all neutrino species with $M_{\nu i} \le 1$ MeV. Observational data imply that $\Omega_m h^2 \approx 0.1 - 0.3$ since $\Omega_m \approx 0.3 - 0.5$ and $h \simeq 0.65 \pm 0.1$. Thus if all the dark matter were light neutrinos, the sum of their masses would be $\approx 9 - 28$ eV.

In deriving above equation, we have been assuming that all the neutrino species are light enough to still be relativistic at decoupling, i.e. lighter than an MeV. The bound shows that they must then be much lighter than that. In the alternative case that a neutrino species is nonrelativistic at decoupling, it has been shown that its mass must then exceed several GeV, which is not true of the known neutrinos $(\nu_e, \nu_\mu$ and $\nu_\tau)$. (One might at first think that the Boltzmann factor would sufficiently suppress the number density of neutrinos weighing a few tens of MeV to allow compatibility with the present density of the universe. It is the fact that they "freeze out" of equilibrium well before the temperature drops to their mass that leads to the higher mass limit.) We have also been assuming that the neutrino chemical potential is negligible, i.e. that $|\mathrm{n}_\nu - n_{\bar{\nu}}| \ll n_\gamma$. This is very plausible, since the net baryon number density $(n_b - n_{\bar{b}}) \le 10^{-9}\ n_\gamma$, and big bang nucleosynthesis restricts the allowed parameters,

Phase Space Constraint

We have just seen that light neutrinos must satisfy an upper bound on the sum of their masses. But now we will discuss a lower bound on neutrino mass that arises because they must be rather massive to form the dark matter in galaxies, since their phase space density is limited by the Pauli exclusion principle. A slightly stronger bound follows from the fact that they were not degenerate in the early universe.

The phase space constraint follows from Jeans's theorem in classical mechanics to the effect that the maximum 6-dimensional phase space density cannot increase as a system of collisionless particles evolves. At early times, before density inhomogenitites become nonlinear, the neutrino phase space density is given by the Fermi-Dirac distribution,

$$n_\nu(p) = \frac{g_\nu}{h^3}\left[1 + \exp\left(\frac{pc}{kT_\nu(z)}\right)\right]^{-1},$$

where here h is Planck's constant and $g_\nu = 2$ for each species of left-handed ν plus right-handed $\bar{\nu}$. Since momentum and temperature both scale as redshift z as the universe expands, this distribution remains valid after neutrinos drop out of thermal equilibrium at ~ 1 MeV, and even into the nonrelativistic regime $T_\nu < m_\nu$. The standard version of the phase space constraint follows from demanding that the central phase space density $9[2(2\pi)^{5/2}Gr_c^2\sigma m_\nu^4]^{1}$ of the DM halo, assumed to be an isothermal sphere of core radius rc and one-dimensional velocity dispersion σ, not exceed the maximum value of the initial phase space density $n_\nu(0) = g_\nu/2h^3$. The result is,

$$m_\nu > (120eV)\left(\frac{100\,\mathrm{km}\,s^{-1}}{\sigma}\right)^{1/4}\left(\frac{1\,\mathrm{kpc}}{r_c}\right)^{1/2}\left(\frac{g_\nu}{2}\right)^{-1/4}.$$

The strongest lower limits on mν follow from applying this to the smallest galaxies. Both theoretical arguments regarding the dwarf spheroidal (dS) satellite galaxies of the Milky Way and data on Draco, Carina, and Ursa Minor made it clear some time ago that dark matter dominates the gravitational potential of these dS galaxies, and the case has only strengthened with time. The phase space constraint then sets a lower limit $m_\nu > 500$ eV, which is completely incompatible with the cosmological constraint. However, this argument only excludes neutrinos as the DM in certain small galaxies; it remains possible that the DM in these galaxies is (say) baryonic, while that in larger galaxies such as our own is (at least partly) light neutrinos. A more conservative phase space constraint was obtained for the Draco and Ursa Minor dwarf spheroidals, but the authors concluded that neutrinos consistent with the cosmological upper bound on mν cannot be the DM in these galaxies. A similar analysis applied to the gas-rich lowrotation-velocity dwarf irregular galaxy DDO 154 gave a limit mν > 94 eV, again inconsistent with the cosmological upper bound.

Free Streaming

The most salient feature of hot DM is the erasure of small fluctuations by free streaming. Thus even collisionless particles effectively exhibit a Jeans mass. It is easy to see that the minimum mass of a surviving fluctuation is of order M_{Pl}^3/m_ν^2. Let us suppose that some process in the very early universe — for example, thermal fluctuations subsequently vastly inflated in the inflationary scenario — gave rise to adiabatic fluctuations on all scales. In adiabatic fluctuations, all the components — radiation and matter — fluctuate together. Neutrinos of nonzero mass mν stream relativistically from decoupling until the temperature drops to $T \sim m_\nu$, during which time they traverse a distance $d_\nu = R_H\,(T = m_\nu) \sim M_{Pl}\,m_\nu^{-2}$. In order to survive this free streaming, a neutrino fluctuation must be larger in linear dimension than dv. Correspondingly, the minimum mass in neutrinos of a surviving fluctuation is $M_{J,\nu} \sim d_\nu^3 m_\nu n_\nu (T = m_\nu) \sim d_\nu^3 m_\nu^4 \sim M_{Pl}^3 m_\nu^{-2}$. By analogy with Jeans's calculation of the minimum mass of an ordinary fluid perturbation for which gravity can overcome pressure, this is referred to as the (free-streaming) Jeans mass.

A more careful calculation gives,

$$d_\nu = 41(m_\nu / 30\,\mathrm{eV})^{-1}(1+z)^{-1}\,\mathrm{Mpc},$$

that is, $d_\nu = 41(m_\nu / 30\,\mathrm{eV})^{-1}$ Mpc in comoving coordinates, and correspondingly,

$$M_{J,v} = 1.77 M_{Pl}^3 m_v^{-2} = 3.2 \times 10^{15} (m_v / 30\,\text{eV})^{-2} M_\odot,$$

which is the mass scale of superclusters. Objects of this size are the first to form in a ν-dominated universe, and smaller scale structures such as galaxies can form only after the initial collapse of supercluster-size fluctuations. When a fluctuation of total mass ~ 10^{15} $M_\odot$ enters the horizon at $z \sim 10^4$, the density contrast δ_{RB} of the radiation plus baryons ceases growing and instead starts oscillating as an acoustic wave, while that of the massive neutrinos δ_v continues to grow linearly with the scale factor $R = (1 + z)^{-1}$ since the Compton drag that prevents growth of δRB does not affect the neutrinos.

By recombination, at $Z_r \sim 10^3$, $\delta_{RB}/\delta_v \leq \sim 10^{-1}$, with possible additional suppression of δRB by Silk damping. Thus the hot DM scheme with adiabatic primordial fluctuations predicts small-angle fluctuations in the microwave background radiation that are lower than in the adiabatic baryonic cosmology, which was one of the reasons HDM appealed to Zel'dovich and other theorists. Similar considerations apply in the warm and cold DM schemes. However, as we will discuss in a moment, the HDM top-down sequence of cosmogony is wrong, and with the COBE normalization hardly any structure would form by the present.

In numerical simulations of dissipationless gravitational clustering starting with a fluctuation spectrum appropriately peaked at λ ~ dv (reflecting damping by free streaming below that size and less time for growth of the fluctuation amplitude above it), the regions of high density form a network of filaments, with the highest densities occurring at the intersections and with voids in between. The similarity of these features to those seen in observations was cited as evidence in favor of HDM.

Problems with ν DM

A number of potential problems with the neutrino dominated universe had emerged by about 1983, however.

- From studies both of nonlinear clustering (comoving length scale λ ≤ 10 Mpc) and of streaming velocities in the linear regime (λ > 10 Mpc), it follows that supercluster collapse must have occurred recently: $z_{sc} \leq 0.5$ is indicated and in any case $z_{sc} < 2$. However, the best limits on galaxy ages coming from globular clusters and other stellar populations indicated that galaxy formation took place before z ≈ 3. Moreover, if quasars are associated with galaxies, as is suggested by the detection of galactic luminosity around nearby quasars and the apparent association of more distant quasars with galaxy clusters, the abundance of quasars at $z > 2$ was also inconsistent with the "top-down" neutrino dominated scheme in which superclusters form first: $z_{sc} > z_{galaxies}$.
- Numerical simulations of the nonlinear "pancake" collapse taking into account dissipation of the baryonic matter showed that at least 85% of the baryons are so heated by the associated shock that they remain unable to condense, attract neutrino halos, and eventually form galaxies. This was a problem for the hot DM scheme for two reasons. With the primordial nucleosynthesis constraint Ωb <≤ 0.1, there would be difficulty having enough baryonic matter condense to form the luminosity that we actually observe. And, where are the X-rays from the shock-heated pancakes.

- The neutrino picture predicts that there should be a factor of ~ 5 increase in M/M_b between large galaxies ($M \sim 10^{12}\ M_\odot$) and large clusters ($M \geq 10^{14} M_\odot$), since the larger clusters, with their higher escape velocities, are able to trap a considerably larger fraction of the neutrinos. Although there is some indication that the mass-to-light ratio M/L increases with M, the ratio of total to luminous mass M/Mlum is probably a better indicator of the value of M/Mb, and it is roughly the same for galaxies with large halos and for rich clusters.

These problems, while serious, would perhaps not have been fatal for the hot DM scheme. But an even more serious problem for HDM arose from the low amplitude of the CMB fluctuations detected by the COBE satellite, $(\Delta T / T)_{rms} = (1.1 \pm 0.2) \times 10^{-5}$ smoothed on an angular scale of about 10°. Although HDM and CDM both have the Zel'dovich spectrum shape ($P(k) \propto k$) in the long-wavelength limit, because of the free-streaming cutoff the amplitude of the HDM spectrum must be considerably higher in order to form any structure by the present. With the COBE normalization, the HDM spectrum is only beginning to reach nonlinearity at the present epoch. Thus the evidence against standard hot DM is convincing. At very least, it indicates that structure formation in a neutrino-dominated universe must be rather more complicated than in the standard inflationary picture. The main alternative that has been considered is cosmic strings plus hot dark matter. Because the strings would continue to seed structure up until the present, and because these seeds are in the nature of rather localized fluctuations, hot DM would probably work better with string seeds than cold DM. However, strings and other cosmic defect models are now essentially ruled out because they predict that the cosmic microwave background would have an angular power spectrum without the pronounced (doppler/acoustic/Sakharov) peak at angular wavenumber $l \sim 220$ that now appears to be clearly indicated by the data, along with secondary peaks at higher l.

Mixed Dark Matter

Mixed dark matter, i.e. a mixture of cold and hot dark matter, became fashionable for a few years in the mid-1990. The main virtue of this scenario was that it could reconcile the CMBR observations of the time with (apparently spurious) reports of neutrino masses in the 20–30 eV range. Interest in mixed dark matter declined, however, once the LCDM model entered the stage and provided a superior explanation for supernova type Ia data, the CMBR anisotropies and the observed large scale structure of the Universe at both high and low redshift. While neutrinos are no longer believed to contribute substantially to the energy density of the Universe, they do appear to have non-zero masses in the right range to make them act as HDM. A small contribution from HDM (with $\Omega_{HDM} \sim 0.01$) is still viable within the LCDM picture, and could help to lower the central densitites of dark halos somewhat.

The Lyman$-\alpha$ forest constraint on the WDM mass in MDM models is milder than that in pure WDM models . The structure formation in MDM models is not well established even after some previous efforts. MDM models also attract interests in the context of the reported anomalous X-ray line in stacked X-ray spectra in XMM-Newton and Chandra data. While the anomaly has not been confirmed in Suzaku data , the 3.5 keV unidentified X-ray line may originate from the decay of sterile neutrinos. Harada and Kamada show that decaying 7 keV sterile neutrinos that are produced via non-resonant process called Dodelson-Widrow mechanism can reproduce the 3.5 keV X-ray line if they account for 20–60% of the present mass density of DM. Interestingly this MDM model can also mitigate the missing satellite problem while evading constraints from the Lyman $-\alpha$ forests.

Dark Matter Halo

A dark halo is the inferred halo of invisible material (dark matter) that permeates and surrounds individual galaxies, as well as groups and clusters of galaxies.

Evidence for the existence of dark matter first came from studies of the motions of stars and gas in galaxies. For example, the quantity and distribution of luminous matter within disk galaxies cannot account for the rotation curves observed, implying a significant invisible component. In a similar manner, velocity dispersions measured in the outer regions of elliptical galaxies are higher than expected given the luminous matter within the galaxy. Estimates based on these considerations suggest that perhaps as much as 90% of the matter present in galaxies is in the form of dark matter.

Evidence for dark matter is also found through observations of the motions of galaxies in groups and clusters. Using a similar argument to that applied to stellar motions within galaxies, the velocities of galaxies in groups and clusters are so high that the group or cluster would fly apart if only the luminous matter were present.

In addition, the X-ray emitting gas often found throughout galaxy clusters indicates that these clusters must contain large amounts of dark matter. With a temperature in excess of a million degrees, this gas would evaporate from the cluster if the visible elements were the only components. In both galaxies and groups and clusters of galaxies, the dark matter is found to be distributed in a roughly spherical halo around the visible component – the dark halo. In the Milky Way the dark halo appears to extend out to at least 300,000 light years, and possibly even further, reaching far beyond the extent of the visible matter in the disk.

Rotation Curves as Evidence of a Dark Matter Halo

The presence of dark matter (DM) in the halo is inferred from its gravitational effect on a spiral galaxy's rotation curve. Without large amounts of mass throughout the (roughly spherical) halo, the rotational velocity of the galaxy would decrease at large distances from the galactic center, just as the orbital speeds of the outer planets decrease with distance from the Sun. However, observations of spiral galaxies, particularly radio observations of line emission from neutral atomic hydrogen (known, in astronomical parlance, as HI), show that the rotation curve of most spiral galaxies flattens out, meaning that rotational velocities do not decrease with distance from the galactic center. The absence of any visible matter to account for these observations implies either that unobserved ("dark") matter, first proposed by Ken Freeman in 1970, exist, or that the theory of motion under gravity (General Relativity) is incomplete. Freeman noticed that the expected decline in velocity was not present in NGC 300 nor M33, and considered an undetected mass to explain it. The DM Hypothesis has been reinforced by several studies.

Formation and Structure of Dark Matter Halos

The formation of dark matter halos is believed to have played a major role in the early formation of galaxies. During initial galactic formation, the temperature of the baryonic matter should have still been much too high for it to form gravitationally self-bound objects, thus requiring the prior formation of dark matter structure to add additional gravitational interactions. The current hypothesis for this is based on cold dark matter (CDM) and its formation into structure early in the universe.

The hypothesis for CDM structure formation begins with density perturbations in the Universe that grow linearly until they reach a critical density, after which they would stop expanding and collapse to form gravitationally bound dark matter halos. These halos would continue to grow in mass (and size), either through accretion of material from their immediate neighborhood, or by merging with other halos. Numerical simulations of CDM structure formation have been found to proceed as follows: A small volume with small perturbations initially expands with the expansion of the Universe. As time proceeds, small-scale perturbations grow and collapse to form small halos. At a later stage, these small halos merge to form a single virialized dark matter halo with an ellipsoidal shape, which reveals some substructure in the form of dark matter sub-halos.

The use of CDM overcomes issues associated with the normal baryonic matter because it removes most of the thermal and radiative pressures that were preventing the collapse of the baryonic matter. The fact that the dark matter is cold compared to the baryonic matter allows the DM to form these initial, gravitationally bound clumps. Once these subhalos formed, their gravitational interaction with baryonic matter is enough to overcome the thermal energy, and allow it to collapse into the first stars and galaxies. Simulations of this early galaxy formation matches the structure observed by galactic surveys as well as observation of the Cosmic Microwave Background.

Density Profiles

A commonly used model for galactic dark matter halos is the pseudo-isothermal halo:

$$\rho(r) = \rho_o \left[1 + \left(\frac{r}{r_c} \right)^2 \right]^{-1}$$

where ρ_o denotes the finite central density and r_c the core radius. This provides a good fit to most rotation curve data. However, it cannot be a complete description, as the enclosed mass fails to converge to a finite value as the radius tends to infinity. The isothermal model is, at best, an approximation. Many effects may cause deviations from the profile predicted by this simple model. For example, (i) collapse may never reach an equilibrium state in the outer region of a dark matter halo, (ii) non-radial motion may be important, and (iii) mergers associated with the (hierarchical) formation of a halo may render the spherical-collapse model invalid.

Numerical simulations of structure formation in an expanding universe lead to the theoretical prediction of the NFW (Navarro-Frenk-White) profile:

$$\rho(r) = \frac{\rho_{crit}\delta_c}{\left(\frac{r}{r_s}\right)\left(1+\frac{r}{r_s}\right)^2}$$

where r_s is a scale radius, δ_c is a characteristic (dimensionless) density, and $\rho_{crit} = 3H^2 / 8\pi G$ is the critical density for closure. The NFW profile is called 'universal' because it works for a large variety of halo masses, spanning four orders of magnitude, from individual galaxies to the halos of galaxy clusters. This profile has a finite gravitational potential even though the integrated mass still diverges logarithmically. It has become conventional to refer to the mass of a halo at a fiducial point that encloses an overdensity 200 times greater than the critical density of the universe, though mathematically the profile extends beyond this notational point. It was later deduced that the density profile depends on the environment, with the NFW appropriate only for isolated halos. NFW halos generally provide a worse description of galaxy data than does the pseudo-isothermal profile, leading to the cuspy halo problem.

Higher resolution computer simulations are better described by the Einasto profile:

$$\rho(r) = \rho_e Exp\left[-d_n\left(\left(\frac{r}{r_c}\right)^{\frac{1}{n}} - 1\right)\right]$$

where r is the spatial (i.e., not projected) radius. The term d_n is a function of n such that ρ_e is the density at the radius r_e that defines a volume containing half of the total mass. While the addition of a third parameter provides a slightly improved description of the results from numerical simulations, it is not observationally distinguishable from the 2 parameter NFW halo, and does nothing to alleviate the cuspy halo problem.

Shape

The collapse of overdensities in the cosmic density field is generally aspherical. So, there is no reason to expect the resulting halos to be spherical. Even the earliest simulations of structure formation in a CDM universe emphasized that the halos are substantially flattened. Subsequent work has shown that halo equidensity surfaces can be described by ellipsoids characterized by the lengths of their axes.

Because of uncertainties in both the data and the model predictions, it is still unclear whether the halo shapes inferred from observations are consistent with the predictions of ΛCDM cosmology.

Halo Substructure

Up until the end of the 1990s, numerical simulations of halo formation revealed little substructure. With increasing computing power and better algorithms, it became possible to use greater numbers of particles and obtain better resolution. Substantial amounts of substructure are now expected. When a small halo merges with a significantly larger halo it becomes a subhalo orbiting within the potential well of its host. As it orbits, it is subjected to strong tidal forces from the host, which cause it to lose mass. In addition the orbit itself evolves as the subhalo is subjected to dynamical friction which causes it to lose energy and angular momentum to the dark matter particles of its host. Whether a subhalo survives as a self-bound entity depends on its mass, density profile, and its orbit.

Angular Momentum

As originally pointed out by Hoyle and first demonstrated using numerical simulations by Efstathiou & Jones, asymmetric collapse in an expanding universe produces objects with significant angular momentum.

Numerical simulations have shown that the spin parameter distribution for halos formed by dissipation-less hierarchical clustering is well fit by a log-normal distribution, the median and width of which depend only weakly on halo mass, redshift, and cosmology:

$$\rho(\lambda)d\lambda = \frac{1}{\sqrt{2\pi}\sigma_{ln\lambda}}\exp\left[-\frac{\ln\left(\frac{\lambda}{\bar{\lambda}}\right)^2}{2\sigma^2_{\ln\lambda}}\right]\frac{d\lambda}{\lambda}$$

with $\bar{\lambda} \approx 0.035$ and $\sigma_{ln\lambda} \approx 0.5$. At all halo masses, there is a marked tendency for halos with higher spin to be in denser regions and thus to be more strongly clustered.

Theories about the Nature of Dark Matter

The nature of dark matter in the galactic halos of spiral galaxies is still undetermined, but there are two popular theories: either the halo is composed of weakly interacting elementary particles known as WIMPs, or it is composed of a number of small, dark bodies known as MACHOs. MACHOs, an acronym for massive compact halo objects, would be composed of "ordinary" matter that simply does not emit easily detectable radiation. The reliance of modern astronomy on detecting electromagnetic radiation would render dim, but massive objects, nearly undetectable. A wide range of possible MACHO candidates have been suggested, encompassing everything from black holes to very dim dwarf stars. Some have hypothesised that the dark matter halo consists of primordial black holes. Despite being too dim to detect directly via electromagnetic telescopes, MACHOs would necessarily interact gravitationally, as described by general relativity. The preferred method for searching for MACHOs in the halo of our own galaxy has been to look for microlensing events. Gravitational microlensing occurs when two stars fall on a common line of sight, rendering the far star obscured by the near one. However, as light from the far star passes through the gravitational well of the near star the light bends, creating an Einstein Halo. In a microlensing event, the halo is so small that it is optically indistinguishable from the star. The overall effect is to simply make the star appear brighter. The EROS and MACHO projects search of MACHOs in the halo by observing stars in the small and large Magellanic clouds. If a MACHO existed in the halo along the line of sight of stars in the cloud, the lensing would brighten it from that orientation as opposed to others. The magnitude and number (or lack of) lensing events can be used to place bounds on the masses of any MACHOs which might be in the halo. The two projects initially were able to place a very strict limit on MACHOS in the range of $3.5 \cdot 10^{-7} M_{\odot} < m < 4.5 \cdot 10^{-5} M_{\odot}$, concluding that such light mass MACHOs could, at most, only constitute 10% of the accepted mass of our dark matter halo. Two years later, the EROS2 project extended this limit, concluding that any MACHO less than a single solar mass could not make up a significant part of the halo. The two collaborations were able to extend this to a rather large window, ruling out any MACHO in the $3.5^{-7} M_{\odot} < m < 30 M_{\odot}$ window. The final window of extremely heavy MACHOs above $43 M_{\odot}$ were ruled out by comparing Monte Carlo method simulations to observed distributions, which the authors indicate as "The End of the MACHO Era". Extremely light MACHOs less than the current limit are also not a viable possibility, as such light mass objects would not have survived on the timescale that it would take for the galaxy to form.

Dark Halo Around the Milky Way

The Milky Way's dark halo is believed to outweigh the galaxy's normal matter by around a factor of 20. While the inner edge of the luminous hypothesized outer ring that surrounds the spiral disk of the Milky Way may be around 120,000 light-years (ly) across, the dark halo encompasses and permeates even the enormous luminous halo of scattered individual stars and globular clusters, extending some 300,000 to 400,000 ly out from the galactic center in radius (twice that in diameter). In 2006, a team of scientists (including Jürg Diemand, Michael Kuhlen, and Piero Madau) modelled the process by which dark matter "clouds" are attracted to form the Milky Way's dark halo (Diemand et al, 2006). They simulated the development and movement of 234 million "cloudlets" as they come together to form a dark halo about the same size as that around the Milky Way. Their simulations show that there should be at least 10,000 separate "subhaloes" of dark matter within the overall galactic halo, each at least a few thousand light years across. Over time, a fair number of these galactic "seeds" should have attracted ordinary matter (mostly hydrogen and helium gas) to form star clusters. About 120 of the larger clumps of dark matter should have become large enough to have attracted enough gas to become dwarf galaxies, altough astronomers have identified about 15 dwarf satellite galaxies around the Milky Way thus far.

The Milky Way's bright ordinary matter is composed of a spiral disk and outer ring, which in turn is embedded in a larger luminous halo that is only the visible part of an even larger halo of dark matter.

It is possible that many subhaloes did not form dwarf galaxies because dark matter has some property that prevents it from forming dense clumps. For example, it might be unexpectedly hot, and therefore hard to compress. Many astronomers currently believe, however, that there are other explanations for the paucity of observed satellite companions around the Milky Way. It may be that most of the subhaloes were sterilized by ultraviolet light from the earliest stars, which heated up intergalactic gas so that that it has been more difficult for subhaloes to capture. In addition, supernova explosions may have blasted gas out of many early dwarf galaxies, halting their continued development. In November 2009, two astronomers revealed supporting evidence that Smith's Cloud may be surrounded by a massive dark halo.

Smith's Cloud, a massive gas cloud that is apparently surrounded by an even more massive dark halo, will collide with the disk of the Milky Way galaxy within 20-40 million years.

Astronomers have detected evidence that dwarf satellite galaxies are disturbing the cocoon of dark matter around the Milky Way and causing its disk to warp. In 1957, astronomers surveying galactic hydrogen gas discovered that the Milky Way is not flat but warped near its edges like a fedora hat, with one side of its spiral disk curving as much as 20,000 light years above the main galactic plane and the other dipping a little less below it. While some researchers suspected the warp was caused by the two Magellanic Clouds (nearby satellite galaxies that orbit the Milky Way every 1.5 billion years), subsequent calculations showed they alone were not massive enough to produce the disk's warp. In January 2006, a team of researchers (including Leo Blitz, Martin Weinberg, Carl Heiles, and Evan Levine) announced finding evidence that the Magellanic Clouds can account for the warp but only because their motion around the Milky Way generates a powerful gravitational wake within the massive dark halo. As the Magellanic Clouds orbit the Milky Way, computer simulations indicate that the galactic disk ripples over time and its edges ruffle "like a table cloth in the breeze". On January 9, 2007, astronomers announced that new measurements of the velocities of the Megallanic Clouds through space suggest that the Milky Way's combined dark and visible mass must be twice as much as originally thought if the Clouds are trully orbiting satellites of the galaxy.

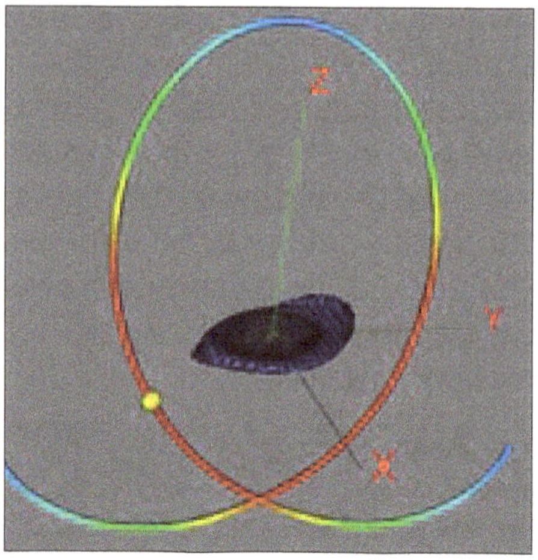

The Milky Way's disk is warped from the orbital movement and interaction of the Magellanic Clouds through the galaxy's dark matter halo.

In January 2010, a team of astronomers (including David Law) announced at the 215th Meeting of the American Astronomical Society that the cloud of dark matter that surrounds our Milky Way galaxy appears to be shaped like a "squashed beach ball" that is oriented perpendicularly to the galaxy's spiral disk. The team studied the path of a dwarf galaxy called the Sagittarius dwarf elliptical or spheroidal galaxy, whose stars have been "shredded" into a long tidal stream as the smaller galaxy was gravitationally drawn into a merger with the Milky Way beginning some three billion years ago. Their hypothesis is that the the gravitional pull of the Milky Way's immense halo of surrounding dark matter should have shaped the trajectory of the tidal stream of stars ripped from the Sagittarius galaxy as it was drawn in.

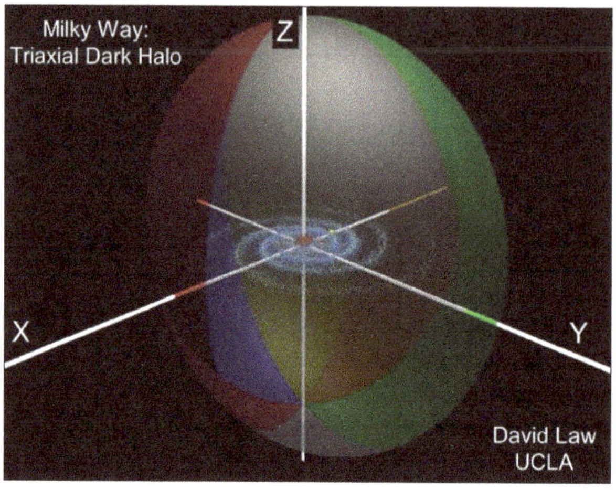

The Milky Way's halo of dark matter is oriented roughly perpendicularly to the spiral disk in the shape of a squashed beach ball -- Sol's location is markeded by the yellow dot.

Their study of this "debris" stream of stars suggests that the distribution of dark matter around the Milky Way is very different to that of the galaxy's stars and gas matter seen in luminous ordinary matter. Although computer simulations had suggested that the halo should mimick the Milky Way's spiral disk of stars, the team's results indicate that the dark halo is oriented roughly perpendicular to the disk and is distributed roughly half as thick as it is wide.

The "tidal stream" of a small galaxy merging into the Milky Way (SagDEG), detected using red (M) giant stars from the Two-Micron All Sky Survey (2MASS), was analyzed to suggest the shape of the Milky Way's halo of dark matter.

References

- Freese, Katherine (4 May 2014). The Cosmic Cocktail: Three Parts Dark Matter. Princeton University Press. ISBN 978-1-4008-5007-5
- Bergstrom, L. (2000). "Non-baryonic dark matter: Observational evidence and detection methods". Reports on Progress in Physics. 63 (5): 793–841. arXiv:hep-ph/0002126. Bibcode:2000RPPh...63..793B. doi:10.1088/0034-4885/63/5/2r3
- Overbye, Dennis (December 27, 2016). "Vera Rubin, 88, Dies; Opened Doors in Astronomy, and for Women". The New York Times. Retrieved 04, April 2020
- Mark H. Jones; Robert J. Lambourne; David John Adams (2004). An Introduction to Galaxies and Cosmology. Cambridge University Press. p. 21; Figure 1.13. ISBN 0-521-54623-0
- Theo Koupelis; Karl F Kuhn (2007). In Quest of the Universe. Jones & Bartlett Publishers. p. 492; Figure 16-13. ISBN 0-7637-4387-9
- Stonebraker, Alan (3 January 2014). "Synopsis: Dark-Matter Wind Sways through the Seasons". Physics – Synopses. American Physical Society. Retrieved 17, August 2020
- Some details of Zwicky's calculation and of more modern values are given in Richmond, M., Using the virial theorem: the mass of a cluster of galaxies, retrieved 28, June 2020

Permissions

All chapters in this book are published with permission under the Creative Commons Attribution Share Alike License or equivalent. Every chapter published in this book has been scrutinized by our experts. Their significance has been extensively debated. The topics covered herein carry significant information for a comprehensive understanding. They may even be implemented as practical applications or may be referred to as a beginning point for further studies.

We would like to thank the editorial team for lending their expertise to make the book truly unique. They have played a crucial role in the development of this book. Without their invaluable contributions this book wouldn't have been possible. They have made vital efforts to compile up to date information on the varied aspects of this subject to make this book a valuable addition to the collection of many professionals and students.

This book was conceptualized with the vision of imparting up-to-date and integrated information in this field. To ensure the same, a matchless editorial board was set up. Every individual on the board went through rigorous rounds of assessment to prove their worth. After which they invested a large part of their time researching and compiling the most relevant data for our readers.

The editorial board has been involved in producing this book since its inception. They have spent rigorous hours researching and exploring the diverse topics which have resulted in the successful publishing of this book. They have passed on their knowledge of decades through this book. To expedite this challenging task, the publisher supported the team at every step. A small team of assistant editors was also appointed to further simplify the editing procedure and attain best results for the readers.

Apart from the editorial board, the designing team has also invested a significant amount of their time in understanding the subject and creating the most relevant covers. They scrutinized every image to scout for the most suitable representation of the subject and create an appropriate cover for the book.

The publishing team has been an ardent support to the editorial, designing and production team. Their endless efforts to recruit the best for this project, has resulted in the accomplishment of this book. They are a veteran in the field of academics and their pool of knowledge is as vast as their experience in printing. Their expertise and guidance has proved useful at every step. Their uncompromising quality standards have made this book an exceptional effort. Their encouragement from time to time has been an inspiration for everyone.

The publisher and the editorial board hope that this book will prove to be a valuable piece of knowledge for students, practitioners and scholars across the globe.

Index

N

P

Q

R

S

T

U

V

W

www.ingramcontent.com/pod-product-compliance
Lightning Source LLC
LaVergne TN
LVHW080718210926
848485LV00003B/488